WORLD BANK TECHNICAL PAPER NUMBER 154

Environmental Assessment Sourcebook

Volume III
Guidelines for Environmental Assessment of Energy and Industry Projects

Environment Department

The World Bank
Washington, D.C.

Copyright © 1991
The International Bank for Reconstruction
and Development/THE WORLD BANK
1818 H Street, N.W.
Washington, D.C. 20433, U.S.A.

All rights reserved
Manufactured in the United States of America
First printing October 1991
Second printing January 1992

Technical Papers are published to communicate the results of the Bank's work to the development community with the least possible delay. The typescript of this paper therefore has not been prepared in accordance with the procedures appropriate to formal printed texts, and the World Bank accepts no responsibility for errors.

The findings, interpretations, and conclusions expressed in this paper are entirely those of the author(s) and should not be attributed in any manner to the World Bank, to its affiliated organizations, or to members of its Board of Executive Directors or the countries they represent. The World Bank does not guarantee the accuracy of the data included in this publication and accepts no responsibility whatsoever for any consequence of their use. Any maps that accompany the text have been prepared solely for the convenience of readers; the designations and presentation of material in them do not imply the expression of any opinion whatsoever on the part of the World Bank, its affiliates, or its Board or member countries concerning the legal status of any country, territory, city, or area or of the authorities thereof or concerning the delimitation of its boundaries or its national affiliation.

The material in this publication is copyrighted. Requests for permission to reproduce portions of it should be sent to Director, Publications Department, at the address shown in the copyright notice above. The World Bank encourages dissemination of its work and will normally give permission promptly and, when the reproduction is for noncommercial purposes, without asking a fee. Permission to photocopy portions for classroom use is not required, though notification of such use having been made will be appreciated.

The complete backlist of publications from the World Bank is shown in the annual *Index of Publications*, which contains an alphabetical title list (with full ordering information) and indexes of subjects, authors, and countries and regions. The latest edition is available free of charge from the Publications Sales Unit, Department F, The World Bank, 1818 H Street, N.W., Washington, D.C. 20433, U.S.A., or from Publications, The World Bank, 66, avenue d'Iéna, 75116 Paris, France.

ISSN: 0253-7494

Library of Congress Cataloging-in-Publication Data
(Revised for vol. 3)

International Bank for Reconstruction and Development.
  Environment Dept.
  Environmental assessment sourcebook.

  (World Bank technical paper, 0253-7494 ;
no. 139)
  Includes bibliographical references.
  Contents: v. 1. Policies, procedures, and
cross-sectoral issues —   — v. 3. Guidelines
for environmental assessment of energy and industry
projects.
    1. Economic development—Environmental aspects.
2. Environmental impact analysis.   I. Title.
II. Series: World Bank technical paper ; no. 139, etc.
TD195.E25I58   1991   333.73'14   91-4324
  ISBN 0-8213-1843-8 (v. 1)
  ISBN 0-8213-1845-4 (v. 3)

Volume III: Guidelines for an Environmental Assessment of Energy and Industry Projects

CONTENTS

Foreword	v
Preface	vii
How to Use the Sourcebook	ix
Acknowledgments	xi

Chapter 10 Energy and Industry — 1

Industrial Hazard Management	2
Hazardous Materials Management	17
Plant Siting and Industrial Estate Development	19
Electric Power Transmission Systems	25
Oil and Gas Pipelines	32
Oil and Gas Development--Offshore	41
Oil and Gas Development--Onshore	52
Hydroelectric Projects	63
Thermoelectric Projects	74
Financing Nuclear Power: Options for the Bank	83
Cement	90
Chemical and Petrochemical	101
Fertilizer	111
Food Processing	120
Small- and Medium-Scale Industries	130
Iron and Steel Manufacturing	135
Nonferrous Metals	145
Petroleum Refining	157
Pulp, Paper, and Timber Processing	168
Mining and Mineral Processing	179

References — 195

Annex 10-1. Sample Terms of Reference (TOR) An Environmental Assessment of Energy Facilities	205
Annex 10-2. Sample Terms of Reference (TOR) An Environmental Assessment of Industrial Facilities	211

Abbreviations/Acronyms — 217

Environmental Assessment - A Guide To Further Reading — 223

FOREWORD

The Sourcebook is designed to assist all those involved in environmental assessment (EA). They include the environmental assessors themselves, project designers and World Bank task managers (TMs). This focus supports an important premise of EA, that sustainable development is achieved most efficiently when negative environmental impacts are identified and addressed at the earliest possible planning stage. The Sourcebook provides practical guidance for designing just such sustainable Bank-assisted projects.

Environmental assessment teams conducting EAs for borrowing governments need to know Bank policy regarding the project under consideration and which aspects of a project are of particular concern to the Bank. This Sourcebook provides specific information and common ground for discussion among those involved: EA professionals, the Bank and borrowing governments.

Project designers need to know applicable Bank requirements and the environmental implications of their design choices. In addition, they need to understand the objectives of an EA team. The Sourcebook provides not only project-specific considerations, but establishes common ground for general discussion, such as that regarding country strategy.

TMs are responsible for ensuring that borrowers fulfill Bank requirements for environmental review, including EAs where indicated. The Sourcebook provides assistance for these advisory tasks, through discussions of fundamental environmental considerations (with emphasis on those with relatively more impact); summaries of relevant Bank policies; and analyses of other topics that affect project implementation (e.g., financial intermediary lending, community involvement, economic evaluation).

Additional audiences likely to be interested in the Sourcebook are other economic development and finance agencies, EA teams for non-Bank projects, environmentalists, academics and NGOs.

The Sourcebook focuses on those operations with major potential for negative environmental impact, such as new infrastructure, dams and highways. Projects with relatively less negative potential, such as maintenance and rehabilitation, are not examined in detail at this stage; they merit a companion volume.

Bank policies and procedures, guidelines, precedents and "best practice" regarding the environment have been scattered throughout the institution and its publications -- or have resided only in the heads of Bank staff. This Sourcebook now collects this corporate knowledge into a single source. It is planned to be an easy-to-use reference manual, hence the overlaps and repetition. Its format is designed to facilitate the frequent updating necessary in a rapidly changing field such as the environment. The Table of Contents is the most efficient entry point from which to locate sections relevant for an individual user. Comments are invited at any stage from users on ways the Sourcebook can better meet their needs.

PREFACE
TO THE FIRST EDITION

This Environmental Sourcebook has been circulating for use as a draft for nearly one year. It is now offered to you for trial use. We seek your assistance on the present contents to alert us to any missing topics that should be included and we request further comments from "users." If at any time in the use of the Sourcebook you have comments, please let us know. The most efficient way to convey your views is to return marked up pages to my office (Room S-5029; or Fax: 202/477-0565).

The Sourcebook will be revised as new information and experience are acquired. The most up-to-date version will be available electronically for all people accessing the Bank's "All-in-one" electronic mail. With continued review and evaluation of experience, we expect to maintain this document as a living, working and up-to-date resource.

Kenneth Piddington
Director
Environment Department
The World Bank
1818 H Street, N.W.
Washington, DC 20433
USA

HOW TO USE THE SOURCEBOOK

The Sourcebook is designed to facilitate the environmental assessment process. It is intended to be used by all involved in EA, primarily the EA practitioner, but also groups managing them, project designers, task managers and environmentalists in general. While much of the document refers mainly to project loans, policy-based and adjustment lending may be addressed subsequently. The Sourcebook is a reference manual which contains the information needed to manage the process of environmental assessment according to the requirements of the World Bank's Operational Directive on EA (OD 4.00, Annex A, October 1989). It is a long document because of the wide range of subjects addressed. However, no one user will need all of the information in the book. Its contents have therefore been organized to be individually as easily accessible as possible, and there is a logical way in which a user can find the items that are pertinent to any particular lending operation. The Sourcebook focuses on operations with more potential for negative environmental impacts, such as major new infrastructure, rather than on operations with less potential impact, such as rehabilitation and maintenance, important though these investments undoubtedly are.

The <u>Table of Contents</u> is the most important section of the Sourcebook. It will assist the user of this reference manual who may be concerned about a specific operation (see the irrigation example below).

Chapter 1 is recommended reading for anyone responsible for a Bank-supported project with potentially significant environmental impacts. It summarizes Bank EA requirements and outlines the Bank's environmental review process, from screening at the time of project identification, right through to post-completion evaluation. A number of "boxes" illustrate different applications of EA in development activities. OD 4.00, Annex A is appended to Chapter 1, along with a list of other Bank operational policy and procedural documents relevant to EA. Annex 1-3 offers a standard format for Terms of Reference (TOR) for an EA that TMs may want to tailor to their specific needs.

Chapters 2 and 3 are "issues" chapters. They provide information and guidance on a number of topics, some of which are likely to arise in any EA. The issues in Chapter 2 are primarily ecological, while those in Chapter 3 are social and cultural. The chapters can, of course, be read in their entirety, but there are two other ways to use them. Their subtopics are shown in the Table of Contents, allowing the user to find them individually. They are also cited where applicable in the discussions of EA guidelines for specific project types, so that they can be referred to in the course of preparing to conduct a particular EA.

Chapters 4, 5, and 6 are "methods" chapters: economics, institutions, and financial intermediary lending. They are not intended to substitute for the knowledge and skills of experts carrying out the actual EA. Chapter 4 gives Sourcebook users an idea of what can be accomplished in the way of economic evaluation of environmental costs and benefits as part of an EA. Chapter 5 addresses institutional strengthening. It stresses the need to develop local capability in EA, identifies some of the broader needs for building country environmental management capacity that an EA may disclose, and considers what may be realistically expected in either area from a single loan or credit. Chapter 6 discusses the particular problems associated with EAs of sector and financial intermediary lending.

The extent to which these chapters are important to an individual user depends on the type of project and the nature of environmental management in the borrowing country.

Chapter 7, community involvement and the role of nongovernmental organizations in EA, explores the implications of OD 4.00, Annex A requirements in this area and offers guidance on how to meet them. Because community involvement is a new concept not only to some Bank staff but also to officials in borrowing countries, the chapter is recommended reading for all environmental assessors as well as task managers.

Chapters 8, 9 and 10 contain sectoral guidelines for EAs. The chapters begin with general considerations pertaining to EA in the sector(s) covered and with discussions of particularly relevant topics (e.g., "Integrated Pest Management and Use of Agrochemicals" in Chapter 8, which concerns the agricultural sector, or plant siting in Chapter 10, on industrial and energy sector projects). The topics are shown in the Table of Contents and cross-referenced throughout the Sourcebook. The balance of each chapter covers specific types of projects, chosen primarily because they have potentially significant environmental impacts. For each type, the project is briefly described (intended only to indicate the features of the project which have environmental significance), potential impacts are summmarized, and special issues are noted that should be considered in an EA. Possible alternatives to the project are outlined, and discussions of management and training needs and monitoring requirements are added. Each review concludes with a table of potential impacts and the measures which can be used to mitigate them. Sample Terms of Reference for the various project types are collected in one section in each chapter.

In the case of a loan for an irrigation project to reclaim arid land, the user would at a minimum consult the following Sourcebook sections:

Chapter 1: "The Environmental Review Process" (if not already acquainted with Bank EAs)

Chapter 8: "Irrigation and Drainage" (for the project-specific guidelines and sample TORs)

Chapter 2: "Arid and Semi-Arid Lands" and "Land and Water Resource Management" (for a review of ecological issues)

Chapter 7: "Community Involvement and the Role of NGOs in Environmental Review" (if not already familiar with the topic in Bank EAs)

The need for other information will become apparent; for example, tribal peoples, international waterways, new land settlement, resettlement, or institutional strenghtening may emerge as important concerns in the project, and the appropriate Sourcebook sections can be consulted.

ACKNOWLEDGMENTS

The Sourcebook staff is indebted to Bank colleagues and consultants who contributed to the first edition of The Environmental Assessment Sourcebook. We wish to express our gratitude to the members of the Environmental Assessment Steering Committee for overseeing this project from beginning to end. We thank our colleagues in the international community as well as colleagues in government and environmental agencies for their comments on various sections of the Sourcebook, and for sharing their own materials.

Environment Department: Kenneth Piddington, Director.

Environmental Assessment Sourcebook Staff: The Environmental Assessment Sourcebook was compiled and edited by Robert Goodland, Thomas E. Walton III, Valerie Edmundson and Charlotte Maxey.

Environmental Assessment Implementation Steering Committee: Gloria J. Davis, Chair, (ASTEN); Cynthia C. Cook (AFTEN); Colin Rees (ASTEN); Martyn J. Riddle (CENDD); J.A. Nicholas Wallis (EDIAR); Bernard Baratz, Stephen F. Lintner (EMTEN); Cesar A. Plaza (LATEN); Surinder P.S. Deol (POPTR); and James Listorti (Consultant).

Chapter 1: The Environmental Review Process: Author: Thomas E. Walton, III (Consultant). Reviewers: Cynthia C. Cook (AFTEN); Walter J. Ochs (AGRPS); Arthur E. Bruestle, Gloria J. Davis and Colin Rees (ASTEN); Thierry Baudon, Stephen F. Lintner, Spyros Margetis and Peter W. Whitford (EMTEN); Robert Goodland (ENVDR); and Albert Printz (Consultant).

Chapter 2: Global and Cross-Sectoral Issues in Environmental Review: Authors: Jan C. Post (ENVAP); Alcira I. Kreimer (ENVPR); Barbara Lausche (LEGOP); Barbara Braatz, Charlotte Maxey, Peter Little, Byron Nickerson, Richard Stoffle, Jon M. Trolldalen, James Talbot and Thomas E. Walton III (Consultants). Reviewers: Agnes Kiss, Walter J. Lusigi, Robert Tillman (AFTEN); David A.P. Butcher, Gloria J. Davis, Colin Rees and Susan S. Shen (ASTEN); Bernard Baratz, Stephen F. Lintner and Peter W. Whitford (EMTEN); Warren D. Fairchild (EMTAG); Robert Goodland (ENVDR); Alcira I. Kreimer (ENVPR); George Ledec (LATEN); Hans J. Peters (INUTD); Albert Printz and Lee Talbot (Consultants).

Chapter 3: Social and Cultural Issues in Environmental Review: Authors: David A.P. Butcher, Gloria J. Davis, Augusta Molnar and William Partridge (ASTEN); Mona Fikri, Jasper Ingersoll, Peter Little, Pam Stambury, Richard Stoffle and June Taboroff (Consultants). Reviewers: Michael M. Cernea, Scott E. Guggenheim (AGRPS); Poul A. Sihm (AFTAG); Cynthia C. Cook, Lee Talbot, Robert Tillman (AFTEN); Valter Angell, Arthur E. Bruestle, Colin Rees (ASTEN); Raymond J. Noronha (ENVAP); Mary B. Dyson and Michael P. Wells (ENVPR); John M. Courtney (ITFPS); Shelton F. Davis (LATEN); and Albert Printz (Consultant).

Chapter 4: Economic Evaluation Methods in Environmental Review: Authors: Herman Daly, Ernst Lutz and Mohan Munasinghe (ENVPR). Reviewers: Valter Angell (ASTEN); Jeremy J. Warford (ENVDR); John Dixon (LATEN); and Albert Printz (Consultant).

Chapter 5: Strengthening Local Capabilities and Institutions: Authors: Gloria J. Davis (ASTEN); Stephen F. Lintner (EMTEN); Barbara Lausche (LEGOP); and Thomas E. Walton III (Consultant). Reviewers: Jean B. Aden (ASTEN); Robert Goodland (ENVDR); Albert Printz (Consultant).

Chapter 6: Sector and Financial Intermediary Lending and Environmental Review: Authors: Martyn J. Riddle (CENDD); Charlotte Maxey and Thomas E. Walton III (Consultants). Reviewers: Rolf Glaeser (AF1IE); Paul Murgatroyd (AS1IE); Jean B. Aden (ASTEN); Kurt M. Constant (ASTIF); Paul A. Popiel (AFTTF); Robert D. Graffam and Rudolf van der Bijl (CCMDR); Millard F. Long (CECFP); Khosrow Zamani (CEMD2); Khalid Siraj (CODOP); Fred D. Levy, Jr. (EAS); Delbert A. Fitchett (EDIAR); Josef Duster (EM1AG); Bernard Baratz (EMTEN); Mark R. Nicholson (INVD1); Samia El Baroudy (LA1TF); and Christophe Bellinger (MIGGU); Melanie Johnson and Albert Printz (Consultants).

Chapter 7: Community Involvement and the Role of NGOs in Environmental Review: Authors: Robert Goodland (ENVDR) and William Nagle (Consultant). Reviewers: Francis J. Lethem (AF2DR); Cynthia C. Cook (AFTEN); Michael M. Cernea, Scott E. Guggenheim (AGRPS); Gloria J. Davis, David A.P. Butcher, William Partridge (ASTEN); Maritta Koch-Weser, Raymond J. Noronha, June Taboroff (ENVAP); Mary Dyson (ENVPR); David M. Beckmann (EXTIE); Shelton H. Davis (LATEN); Nancy Alexander (Friends Committee on National Legislation); Albert Printz (Consultant); and Diane Wood (World Wildlife Fund).

Chapter 8: Agriculture and Rural Development: Authors: Agnes Kiss (AFTEN); Robert Goodland (ENVDR); Anil Somani, Kirk Barker, Susan Braatz, Eugene Dudley, Peter Freeman, John Glenn, Charlotte Maxey, Byron Nickerson, James Talbot, and Thomas E. Walton III (Consultants). Reviewers: Poul A. Sihm (AFTAG); Cynthia C. Cook, Agnes Kiss, Robert Tillman (AFTEN); Guy J.M. LeMoigne (AGRDR); Shawki Barghouti, Walter J. Ochs (AGRPS); John F. Cunningham, Robert G. Grimshaw (ASTAG); Valter Angell, Roger S. Batstone, Susan Braatz, Arthur E. Bruestle, Colin Rees (ASTEN); Martyn J. Riddle (CENDD); J.A. Nicholas Wallis (EDIAR); Warren D. Fairchild, Permanand Gupta, Colin W. Holloway, Gert Van Santen (EMTAG); Bernard Baratz, Anders O. Halldin, Stephen F. Lintner, and Spyros Margetis (EMTEN); Robert J. Goodland (ENVDR); Asif Faiz (INUTD); Daniel Gross (LA1AG); William D. Beattie, Michael J. McGarry (LATAG); Dennis Child (USDA); Albert Printz, James Smyle (Consultants); and David J. Parrish (Virginia Polytechnic Institute and State University).

Chapter 9: Population, Health and Nutrition; Urban Development; Transportation; Water Supply and Sewerage: Authors: Carl R. Bartone (INURD); Albert M. Wright (INUWS); Sandra Cointreau, Colin Franklin, Peter Freeman, James Listorti, Charlotte Maxey, Byron Nickerson, James Talbot and Thomas E. Walton III (Consultants). Reviewers: Cynthia C. Cook, Robert Tillman (AFTEN); Shirin N. Velji (AS2IN); Roger J. Batstone, Arthur E. Bruestle, David G. Williams (ASTEN); Jean H. Doyen (AFTIN); Shirin N. Velji (AS2IN); Martyn J. Riddle (CENDD); A. Amir Al-Khafaji (EM2IN); Mario A. Zelaya (EM3IN); Maurice W. Dickerson (EM4IN); Stephen F. Lintner, Spyros Margetis, Peter W. Whitford (EMTEN); Richard A. MacEwen (EMTIN); Carl R. Bartone, Michael A. Cohen (INURD);

Asif Faiz, Jeffrey S. Gutman, Ian G. Heggie, Hans J. Peters (INUTD); Albert M. Wright (INUWS); John M. Courtney (ITFPS); Shelton H. Davis (LATEN); J. Rausche (United States Army Corps of Engineers); Perry Davies and Albert Printz (Consultants).

Chapter 10: Energy and Industry: Authors: Robert Goodland (ENVDR); Hans Adler, Sandra Cointreau, Eugene Dudley, Valerie Edmundson, Bernanda Flicstein, Ken Kosky, Tom Loomis, John Mulckhuyse, James Newman, Byron Nickerson, Anil Somani, James Talbot, and Thomas E. Walton III (Consultants). Reviewers: Robert Tillman, (AFTEN); John E. Strongman, Peter van der Veen (AFTIE); Uruj Ahmad S. Kirmani, Mihir Mitra, Christopher Wardell (ASTEG); Roger J. Batstone, Colin Rees (ASTEN); Martyn J. Riddle, Jean M.H. Tixhon (CENDD); David A. Craig (EM4IE); Suman Babbar (CFSPS); Bernard Baratz, Anders O. Halldin, and Stephen F. Lintner (EMTEN); Achilles Adamantiades, Mogens H. Fog (EMTIE); Anthony A. Churchill (IENDR); John Homer (IENGU); Alvaro J. Covarrubias, Hernan G. Garcia (LATIE); and Albert Printz (Consultant).

CHAPTER 10

ENERGY AND INDUSTRY

Energy has consistently been the Bank's second largest lending sector in recent years. Within the sector, power projects have accounted for most of the lending. Large hydroelectric and thermoelectric power projects usually have impacts that are varied and potentially very significant. They are often highly controversial as well, from the standpoint of public acceptance. The Sourcebook therefore provides extensive guidance for these types of projects, as well as for power transmission systems.

Loans and credits for oil, gas, and coal development have been a relatively smaller fraction of the sector total. However, because they often involve operations with high potential for pollution or disruption in sensitive natural areas, they are also treated extensively in this chapter.

A large portion of Chapter 10 is devoted to describing the major environmental impacts of typical industrial projects. While lending for industry is normally less than ten percent of annual lending, the types of industrial operations covered in this chapter can have very significant impacts, even when the individual facilities are comparatively small.

In many cases, the project-specific sections of this chapter recommend that pollution control standards for projects be established and enforced. However, prescribing the standards themselves is not a purpose of the Sourcebook. Commonly accepted standards for emissions, effluent discharges, or workplace environmental conditions are presented in some sections, but only as general guidance. They are no substitute for appropriate national standards or, in their absence, standards based on good practice in the particular subsector.

As stated a number of times in the Sourcebook, the Bank views environmental assessment as an opportunity to identify early in the project cycle problems that will become costly when they appear unexpectedly later. In the case of most types of energy and industry investment projects, the ideal point to begin considering environmental impacts is at the time of site selection. An inappropriate site can bring with it environmental and social impacts that will be difficult and expensive to manage at best, whereas through a sound choice of site, some impacts may be avoided altogether. For this reason, a section on plant siting is included in this chapter.

Chapter 10 also contains two other sections of general interest. One concerns management of industrial hazards -- both those that exist in the workplace (which the EA OD includes in the scope of environmental review), and those that occur to the natural surroundings and communities because of the industrial operations. A separate section deals with materials that are of special concern to the Bank, because they either are considered hazardous or are generally known to have the potential to cause significant environmental degradation if mishandled or utilized under conditions that are inappropriate.

INDUSTRIAL HAZARD MANAGEMENT

General Issues

1. Industrial facilities include a wide variety of mining, transportation, energy generation, manufacturing, and waste disposal operations with inherent hazards which require careful management. For example, industrial operations involve the handling, storage, and processing of potentially hazardous substances (e.g., reactive chemicals and hazardous wastes). Industrial facilities also involve potential hazards which are distinct from hazardous substances.

2. Because of the existence of hazards at industrial facilities, the following risks need to be adequately managed to minimize adverse impacts: conditions potentially leading to major release accidents (e.g., releases from pipes, flexible connections, filters, valves, vessels, pumps, compressors, tanks, stacks), occupational health and well-being conditions, and occupational safety conditions.

3. For purposes of this document, hazardous materials and hazardous wastes are categorized by any one or more of the following definitions:

    (a) **Ignitable:** substances which ignite easily and thus pose a fire hazard during normal industrial conditions (e.g., finely divided metals, liquids with flash points of 100°F or lower).

    (b) **Corrosive:** substances which require special containerization because of their ability to corrode standard materials (e.g., acids, acid anhydrides and alkalies).

    (c) **Reactive:** substances which require special storage and handling because they tend to react spontaneously with acid or acid fumes (e.g., cyanides, concentrated alkalies), and because they tend to react vigorously with steam or water (e.g., phosphine, concentrated acids or alkalies), or they tend to be unstable to shock or heat (e.g., pressurized flammable liquids, ordinance supplies) with the result being either the generation of toxic gases, explosion, fire, or the evolution of heat.

    (d) **Toxic:** substances (e.g., heavy metals, pesticides, solvents, petroleum based fuels) which when improperly managed may release toxicants in sufficient quantities to pose a direct chronic or acute health effect through inhalation, skin absorption and ingestion, or lead to a potentially toxic accumulation in the environment and/or food chain.

    (e) **Biological:** substances which when improperly managed may release pathogenic micro-organisms in sufficient quantities to cause infection, pollens, molds, or dander in sufficient quantities to cause allergic responses in those susceptible to the hazard.

4. In addition to the above-described categories of hazardous substances, there are general hazards associated with industrial facilities. These hazards include the following categories:

(a) **Electrical:** electrocution from live conductors and misuse of power tools, overhead power lines, downed electrical wires, buried cables, and work during electric storms.

(b) **Structural:** potential for falling or strain when working conditions include slippery surfaces, steep grades, narrow stairs, open holes, trip hazards and unstable flooring; potential puncture from sharp objects, and potential burial from trench or mine cave-in or from unstable slopes on material stockpiles.

(c) **Mechanical:** collision accidents with moving equipment, especially when operating in reverse, failed pulleys, snapped cables, and clothes catching in gears or drills.

(d) **Temperature:** heat stress in hot environments or when working in clothing which limits the dissipation of body heat and moisture; cold stress in cold environments or when the wind-chill factor is low.

(e) **Noise:** stress and physical damage to the ear when subjected to noise levels exceeding recommended guidelines (e.g., an 8-hour, time-weighted average sound level of 90 dBA, decibels on the A-weighted scale).

(f) **Radiation:** burns and/or internal damage when subjected to excessive levels of ionizing radiation.

(g) **Oxygen Deficiency:** health effects due to displacement of oxygen by another gas or consumption of oxygen by a chemical reaction, particularly in confined spaces or low-lying areas, may occur when levels drop below 19.5 percent oxygen.

5. Ergonomic stresses may result from inappropriately designed tools or work areas which can cause workers to experience discomfort, mental stress, loss of efficiency or loss of well-being. While ergonomic stresses are not hazards in the sense described above, they may lessen a worker's abilities to respond clearly and quickly to a hazard, and thus need to be considered in project development. When stress results from human reaction to monotony, fatigue, repeated movement, or repeated shock, the potential increases for hazards and accidents to occur.

Bank Policy, Procedures, and Guidelines

6. In 1988 the Bank issued Technical Paper 55, <u>Techniques for Assessing Industrial Hazards</u>. This document provides guidelines for identifying potential hazards of major consequence, particularly major release events. This document also provides guidelines for analyzing the magnitude of risk and the potential area of impact (i.e., effect distance or damage range).

7. In the Appendix of <u>Techniques for Assessing Industrial Hazards</u>, the Bank outlines its guidelines for identifying, analyzing and controlling major hazards at industrial installations. As described by the Bank's guidelines, a major hazard would exist under the following circumstances: a release of toxic sub-substances, highly reactive or explosive substances, or flammable substances. The guidelines list the types of industrial installations which could present a major hazard. The guidelines also list substances

and quantities which could present a major hazard. When there is a potential major hazard within a proposed Bank-financed project, Bank policy requires a "Major Hazard Assessment."

8. The Major Hazard Assessment is to be written as part of project preparation. It is to be additional to the Environmental Impact Assessment, and referred to by the Environmental Impact Assessment. The objectives of the Major Hazard Assessment, as outlined in the above-mentioned guidelines, are as follows:

- to identify the nature and the scale of use of dangerous substances at the installation;

- to specify arrangements made for safe operation of the installation for control of serious deviation that could lead to a major accident and for emergency procedures at the site;

- to identify the type, relative likelihood, and broad consequences of major accidents; and

- to demonstrate that the developer has appreciated the major hazard potential of the company's activities and has considered whether the controls are adequate.

9. In 1984 the Bank issued <u>Occupational Health and Safety Guidelines</u>, which review conditions at various categories of its industrial facilities and summarize the main health and safety hazards. These guidelines outline relevant control measures, training, and monitoring. It is Bank policy to require that, at a minimum, the Bank's guidelines for protection of occupational health and safety are to be followed in projects financed by the Bank. It is recommended that a "Health and Safety Plan" be developed as part of project preparation wherever there is risk of either a major release event or significant occupational and health consequences. The contents of such a Health and Safety Plan are outlined herein, under Guidance for Environmental Assessments.

10. In 1988 the Bank issued <u>Environmental Guidelines</u>. The guidelines review, for a wide range of industries, the various waste streams anticipated. Opportunities for waste reuse and recycling, as well as waste minimization, are discussed. Industry-specific issues of occupational health and safety are described. Where specific permissible pollutant discharge levels are provided, it is the Bank's policy that these levels are to serve as minimum standards.

11. In 1989 the Bank issued Technical Paper 93, <u>The Safe Disposal of Hazardous Wastes</u>. The document provides information enabling the classification of hazardous waste and the screening of techniques for hazardous waste management. The document also provides minimum design standards (i.e., for the safe disposal of hazardous waste via secure landfill) which projects would have to address to qualify for Bank financing.

12. The documents discussed above are meant to be upgraded on a regular basis, as the state-of-knowledge on hazard identification and hazard management improves. In referring to these documents, the Bank's Environment Department should be contacted to determine whether more up-to-date guidelines are available. Furthermore, if local regulations differ from the Bank's guidelines, the stricter set of requirements shall prevail in any project financed by the Bank.

Relationship to Bank Investments

13. The issue of industrial hazard management is relevant in energy, industry, mining, pollution control, transportation and agriculture projects.

14. Energy projects may have hazards such as the following: toxic and fire hazards from oil spills or gas leaks, mechanical hazards from drilling rigs, noise hazards around generators, physical hazard from inhalation of dusts from coal ash and oil residues, toxic or corrosive leachate from coal and ash piles, chemicals used in water and wastewater treatment, oxygen depletion in tanks, and electrocution from live conductors.

15. Industry projects may have hazards such as the following: physical hazard from machine moving parts, heat stress from strenuous work near furnaces, noise hazard around machinery, dusts from grinding and sawing, rupture of pressurized vessels, exposure to water and wastewater treatment chemicals, explosion during high speed chemical reactions, and toxic fumes from chemical spills.

16. Mining projects may have hazards such as the following: physical hazard from use of explosives and excavation equipment, dusts from drilling, blasting and crushing, oxygen depletion, and toxic gases in underground mines and cave-ins.

17. Pollution control projects may have hazards such as the following: rupture of pressurized vessels (e.g., chlorine tanks at wastewater treatment plants, pressurized containers in incoming solid waste received at an incinerator), explosion or toxic gas generation from incompatible wastes being mixed, release of dusts and mist containing pathogenic micro-organisms during wastewater and solid waste processing operations, and toxic gases from solid waste disposal.

18. Transportation projects may include facilities which are commonly used for the loading, transport and unloading of hazardous substances. As part of both an environmental impact assessment and a major hazard assessment for a transportation project, the potential for a collision or derailment to occur needs to be reviewed. During such an accident, there is potential for a toxic spill, fire and/or explosion.

19. Agricultural projects and control of pests, such as locust, bring unique problems with the handling and storage, use and disposal of pesticides. In Sub-Saharan Africa, disposal of unused pesticides has been a challenging problem for the donor community.

20. The Bank finances projects with the wide range of potential industrial hazards described above. In the cases of industry, energy or pollution control projects, the increased risk of industrial hazards would directly result from the project. In the cases of transportation projects, risk of an industrial hazard probably would be an indirect impact of the project. Through careful project preparation with strict adherence to Bank guidelines for hazard management, the occurrence of industrial hazards would be minimized and the resulting adverse impacts would be mitigated.

Guidance for Environmental Assessments

21. Many industrial hazards occur as unforeseen accidents caused by inadequate operating and maintenance activities. It is the job of the Environmental Impact Assessment and the Major Hazard Assessment to highlight the potential for such accidents, by anticipating the worst case scenario of events

which might cause them to happen, and to provide management and monitoring plans with the goal of risk minimization (for further discussion, see Table 10.1 at the end of this section).

22. Both the draft Environmental Impact Assessment and the draft Major Hazard Assessment are meant to be conducted in parallel with detailed engineering design of the proposed project, and before design is completely finalized. In this way, any hazards which are identified in the draft assessments can be addressed within the final stages of design and the reduction of impacts reflected in the final assessments.

23. Industrial hazards are minimized and managed through use of engineering controls, administrative controls, personnel protection, occupational health and safety training, health and safety planning, and medical monitoring, as discussed below.

24. Engineering controls include the following design and operational changes:

   (a) **Siting.** Facilities with risk of structural collapse, rupture, fire, or explosion will need to be located in geotechnically stable locations (e.g., minimal risk of seismic activity or subsidence).

   (b) **Buffer Zones.** Based on the nature of the potential hazard (e.g., fireball, toxic gas release, spill), facilities will need to have an appropriately sized buffer zone.

   (c) **Layout Design.** Within an installation with industrial hazards, unit operations will need to be laid out so that incompatible substances are not located within proximity of each other (e.g., substances which would react upon mixing to generate heat, fire, gas, explosion, or violent polymerization). Also, incompatible operations are not to be located within proximity of each other (e.g., welding operations are not to be located near storage of ignitable materials).

   (d) **Resource Substitution.** Within processing operations, substitute a hazardous material with a nonhazardous material. Change the form of the material (e.g., to a gas or a liquid) if the resulting form would be less hazardous (e.g., store toxic gases in a suitable solvent form).

   (e) **Resource Minimization.** Minimize the quantities of hazardous materials used by recovering and recycling them within the process operation. Reduce the inventory of hazardous materials in storage. Use more efficient processing techniques.

   (f) **Process or Storage Modifications.** Store hazardous gas as a refrigerated liquid rather than under pressure. Reduce process temperatures and pressures. Change process methods (e.g., change from spray painting to dip or brush painting).

   (g) **Dust Control.** Dust control measures include spraying water (or water with a wetting agent) at the source of dust dispersion, to minimize the generation of dust. Ventilation, collection and filtration are also effective for dust control. Dusty operations should be isolated and/or contained to the extent possible, especially when the dusts could lead to lung diseases such as silicosis, one the most common occupational diseases in the world and most prevalent at mines, brickyards, glassmaking plants, and sand blasting operations. Occupational asthma is caused by a broad array of chemicals and natural substances, including isocyanates, acid anhydrides, danders, grain dust, cotton dust and wood dust.

(h) **Access Control.** Limitation of personnel to those specifically trained in the work conditions present within a potentially hazardous area, including use of personnel identification, double locks, security services, barriers.

(i) **Labeling.** Complete hazard labeling of all switches, valves, containers, and unit operations. In addition to identifying specific hazardous substances by name, also identify the type of hazard (e.g., toxic, reactive, ignitable, explosive).

(j) **Temperature Control.** Provision of air temperature control may be needed at certain operations in order to avoid heat stress or cold stress. A particularly hot or cold operation may need to be segregated from the others in order to minimize the number of personnel exposed.

(k) **Monitoring.** Monitoring of the environment in the immediate vicinity of potential hazards, as well as at the fence-line of the installation, provides an early warning of a hazard occurring. For example, air quality monitoring for volatile organics, oxygen levels, combustible gas levels, and/or specific air constituents could be conducted on a regular basis using portable equipment or a continuous basis with stationary equipment. Smoke detectors, heat monitors, radiation detectors, as appropriate to the type of installation, are used to signal a hazard occurring.

(l) **Shut-Down.** Provide manual and automatic systems for shut-down of electrical systems and/or process operations, so that the release of hazardous material is minimized.

(m) **Secondary Containment.** Provide, as appropriate, systems to contain releases, such as: water curtains to restrict gas release, dikes and portable booms to contain spills, emergency response equipment to collect spilled material, bunkers or blast walls to confine explosions, fire-proofing to limit the spread of fire, absorbents to absorb or adsorb hazardous substance, and buffer zones.

25. Administrative controls are used when it is not possible to reduce exposure to acceptable levels through engineering controls. Administrative controls may include rearranging work schedules to minimize the duration of exposure to hazards and transfer or rotation of personnel who have reached a maximum allowable exposure limit over time.

26. Use of personnel protection equipment is appropriate for work within the vicinity of potential hazards. Personnel protection choices are based on the nature of the hazard, the level and/or concentration of the hazard, the duration of exposure, and the person-specific susceptibility to being adversely affected.

27. When the nature of the hazard is known and routine, the precise type and level of protective gear can be defined and routinely used (e.g., hard hats, chemical-resistant gloves, air-purifying respirators, safety shoes, ear protection, safety glasses). On the other hand, when the nature of the hazard is unknown (e.g., when several hazardous materials accidentally are combined, or when a toxic waste dump is unexpectedly discovered), it may be necessary to use the most conservative type of protective gear (e.g., chemically resistent and gas impermeable suits, self-contained breathing apparatus) -- downgrading only after the hazard is identified as requiring a lower level of protective gear.

28. Personnel protection involves more than special clothing, glasses, hard hats, ear plugs, etc., to protect the body from harm. The following items are also part of personnel protection, as appropriate to the situation: knife (for emergency exit of a protective suit), portable light, personal monitor (e.g., dosimeter for radiation, personal thermometer for heat/cold stress), harness and lifeline, safety belt, two-way radio, locator beacon (e.g., for locating a victim of hazard).

29. Occupational health and safety training is essential to ensure that personnel adhere to appropriate operating practices which minimize adverse health and safety impacts. The following areas of knowledge and experience are considered essential:

   (a) Appreciation of the properties (e.g., flammability, corrosiveness, toxicity, reactivity) of hazardous substances, as well as the levels at which they pose a significant danger requiring protective measures.

   (b) Awareness of early-warning indicators of hazard/risk identification, and ability to recognize potentially hazardous situations.

   (c) Familiarity with engineering controls to avoid occurrence of hazardous situations.

   (d) Familiarity with capabilities and limitations of the facility to response to hazardous emergencies: ventilation systems, plumbing systems, shut-off systems, containment devices, and emergency response procedures as outlined in the appropriate health and safety plans.

   (e) Knowledge of the use and maintenance of emergency response equipment, as well as routine equipment for health and safety monitoring and protection.

   (f) Knowledge of methods and procedures for decontaminating personnel, equipment, and facility, following potential chemical contamination.

   (g) Refresher training and regular drills simulating emergencies and appropriate emergency response procedures.

   (h) Familiarity with and acceptance of the need for continuous reliance on the "Buddy" system. In the Buddy system, work groups are organized so that each employee exposed to hazard is designated to be observed by at least one other employee who would be ready and able to provide immediate emergency assistance as needed.

(i) Empowerment to act decisively in accordance with health and safety plans during potentially hazardous situations or actual emergencies, especially in situations where supervisors are unavailable or have become victims of the emergency.

30. Health and safety planning involves a complete assessment of an installation with all potential hazards identified. The plan provides the following information:

(a) Definition of all potential hazards.

(b) Health and safety implications of each hazard.

(c) Description of routine health and safety management techniques (e.g., health and safety inspections, maintenance/repair follow-up on inspection citations, record-keeping, personnel protective gear, and medical monitoring).

(d) Outline of emergency response procedures following occurrence of a major hazard (e.g., organization structure of key trained personnel to act as emergency responders, action steps for entering and working within zone of hazard, evacuation procedures, protective gear requirements, decontamination procedures, lines of communication, emergency telephone numbers, map of route to nearest emergency medical care).

(e) Follow-up procedures after the emergency is over.

31. In defining potential hazards and the health and safety implications of each hazard, industrialized countries such as the U.S. have regularly updated exposure guidelines (i.e., threshold limit values, called TLV's) based on the current state-of-knowledge. For example, there are time-weighted average threshold limit values (so-called TLV-TWA's) which define the concentration for a normal 8 hour work day, 40 hour work week to which most workers may be exposed without adverse impacts. Similarly, there are short-term exposure limits (so-called TLV-STEL's) which define the maximum concentration to which a worker may be exposed within a 15 minute period without adverse impacts. (There are international telephone numbers for obtaining information about specific chemicals or combinations thereof.)

32. If the hazard involves an area contaminated from a major release of a hazardous material or an area of hazardous waste material, the health and safety plan will need to outline the site control process. Based on knowledge of safe distances relative to site conditions (e.g., wind direction and site topography), site control defines the zones of work for which the corresponding levels of required personnel protection would also be defined (e.g., zone of contamination, zone of decontamination, and support zone).

33. If there is potential that hazardous conditions could extend beyond project site boundaries to properties occupied by residents or farm animals, the plan will need to address emergency notification procedures and possibly evacuation procedures. Early in the health and safety planning stages it will be necessary to designate community coordinators who would be trained to help lead/coordinate emergency

response activities with the community, including conducting training and practice emergency response drills. It is the Bank's policy that the community around a potentially hazardous installation has a right to know the dangers which might occur and what plans have been developed to minimize and manage risk of such dangers occurring.

34. Medical monitoring is necessary for all workers who might be in contact with hazardous substances or hazardous situations. A baseline medical exam conducted at the start of employment defines initial health condition, including blood levels of specific chemicals with which the worker might be working. The baseline examination includes questioning the worker above his/her medical history. Regular exams (e.g., annual) determine whether there have been adverse health effects potentially attributable to the work. It is essential that the medical examiner be adequately trained to recognize sysmptons/signs which might indicate overexposure to hazards to which the worker might be exposed.

Table 10.1. Industrial Hazard Management

Potential Negative Impacts	Mitigating Measures
Direct	
1. Fires, explosions, emission of toxic gases, vapors, dust, emission of toxic liquids, radiation and various combination of these effects.	1. <ul><li>Provision of bunkers or blast walls.</li><li>Firewalls/fireproofing of structures.</li><li>Provision of escape routes for employees.</li><li>Provision of safety and emergency training for employees.</li><li>Implementation of emergency procedure on- and off-site.</li><li>Provision of public alert systems and education of public.</li><li>Planning and training for evacuation.</li><li>Provision of safety buffer zones around the plant boundary.</li></ul>
2. <u>Explosives</u>: explosion	2. <ul><li>Storage and handling should be according to the manufacturers recommendation.</li><li>Special precautions should be taken against theft and fires and during destruction.</li><li>The following general rules should be applied:<ul><li>Lighting in the storage area should be natural or by permissible lights.</li><li>Lamps should be vapor proof and switch should be outside the building.</li><li>Only tools of wood or other non-metallic material should be used.</li><li>Cases of explosive should not be piled in stacks more than 6 feet high.</li></ul></li></ul>

Table 10.1. Industrial Hazard Management (continued)

Potential Negative Impacts	Mitigating Measures
Direct (continued)	
	• Cases of explosives should be stored topside up, so that cartridges are lying flat.
	• They should be turned at regular intervals, as this will help to prevent their deterioration.
3. Flammable Materials: fire hazard	3. • Store in places that are cool enough to prevent accidental ignition in the event that vapors of the flammable materials mix with the air.
	• Provide adequate ventilation in storage space, so that leakage of such vapors from containers will be diluted enough to prevent a spark from igniting them.
	• Locate storage area well away from areas of fire hazard (for example, where torch-cutting of metals is to be performed).
	• Keep apart from powerful oxidizing agents materials that are susceptible to spontaneous heating (explosive or materials that react with air or moisture to evolve heat).
	• Provide fire-fighting equipment.
	• Prevent smoking or use of bare filament heaters.
	• Storage area must be electrically grounded and equipped with automatic smoke or fire detection equipment.
4. Oxidizing Agents: fire hazard	4. • Store away from liquids of low flash point (flammable).
	• Keep area cool and ventilated.

Table 10.1. Industrial Hazard Management (continued)

Potential Negative Impacts	Mitigating Measures
Direct (continued)	
	- Keep fuel away.
	- The area should be fireproof.
	- Note: Normal firefighting equipment is of little use since the blanketing or smothering effect of fire extinguishers is less effective because the oxidizers supply their own oxygen.

5. **Water Sensitive Materials:** evolve heat, flammable gases or explosive gases in contact with water, steam or water solution.	5. - Store in dry and cool areas.
	- Because many of these materials are also flammable, it is essential that no automatic sprinkler system be used in the storage area.
	- Such an area should have no water coming to it at all.
	- Heating may be electrical or with hot, dry air.
	- Storage building must be waterproof, located on high ground and separated from other storage.
	- Particular attention should be paid to the following:
	  • pocketing of light gases under the roof
	  • introduction of sources of ignition
	  • periodic inspection
	  • automatic detection systems
	  • alarms in case of dangerous concentrations of flammable gases

Table 10.1. Industrial Hazard Management (continued)

Potential Negative Impacts	Mitigating Measures
Direct (continued)	
6. <u>Acid and Acid Fume-Sensitive Materials</u>: evolve heat, hydrogen and flammable and/or explosive gases.	6. • Do not store acids in proximity to such materials (e.g., storing acids on structural alloys sheds). • If metal is used in construction, it should be painted or otherwise rendered immune to attack by acid. • Area must be kept cool, ventilated and periodically inspected. • Source of ignition must be kept away.
7. <u>Pressurized Storage of Flammable Fluids</u>: when subjected to fire can cause "Boiling Liquid Expanding Vapor Explosion" (BLEVE).	7. • Tanks should be stored upright and chained or otherwise securely attached to some substantial support to minimize the chance of falling over and breaking or straining the valve or other part of the tank. • Tank storage area should be kept cool, out of direct rays of sun, and away from hot pipes. • Provide means (sprinkler) of keeping the tanks cool in case of external or internal fire. • Take care to keep from damaging tanks in handling. • Valves must be operated carefully and kept in good condition. • Do not hammer valve cocks. • Discourage tampering with tanks in any way.

Table 10.1. Industrial Hazard Management (continued)

Potential Negative Impacts	Mitigating Measures
Direct (continued)	

8. <u>Toxic Materials</u>: cause serious danger (death or serious injury to people or environment).	8. • Reduction of inventories in storage and in process.
	• Modify process or storage conditions (e.g., store and process toxic gases in a large volume of nontoxic carrier material).
	• Store hazardous gas as a refrigerated liquid rather than under pressure.
	• Improve shutdown and secondary containment which will reduce the amount escaping from containment or from site.
	• Automatic shutdown will reduce the amount of material escaping from containment:
	  • water curtains will restrict gas release.
	  • dikes (or bunds) will restrict liquid release.

9. <u>Corrosive Materials</u>: destroy containers, and react to evolve toxic gases in contact with substances such as cyanides and arsenides.	9. • Keep storage or process area cool and ventilated to prevent accumulation of fumes. • Keep containers closed and labeled. • Paint all exposed metal in the vicinity of such storage and check it periodically for weakening by corrosion. • Keep isolated from materials that would produce highly toxic fumes if contacted. • Provide instructions for and supply of specific neutralizing agent to be used in case of spill, leak or major accident.

Table 10.1. Industrial Hazard Management (continued)

Potential Negative Impacts	Mitigating Measures
Indirect	
10. Occupational health effects on workers due to fugitive dust, materials handling, noise, or other process operations. Accidents occur at higher than normal frequency because of level of skill or labor.	10. Facility should implement a Safety and Health Program designed to: • identify, evaluate, monitor, and control health hazards • provide safety training
11. Regional solid waste problem exacerbated by inadequate on-site storage or lack of ultimate disposal facilities.	11. Plan for adequate on-site disposal areas assuming screening for hazardous characteristics of the leachate is known. • Provide, in design phase, for adequate ultimate disposal facilities.
12. Transit patterns disrupted, noise and congestion created, and pedestrian hazards aggravated by heavy trucks transporting raw materials to/from facility.	12. • Site selection can mitigate some of these problems. • Special transportation sector studies should be prepared during project feasibility to select best routes to reduce impacts. • Transporter regulation and development of emergency contingency plans to minimize risk of accidents.

HAZARDOUS MATERIALS MANAGEMENT

1. The Bank's formal position on hazardous waste disposal was stated by the President on 10 July 1988. The international shipment of toxic wastes -- especially from industrial to developing nations for disposal -- poses a threat not only to the environment of the recipient countries, but also to the world's oceans and to the health of the global community.

2. The danger is worsening as producers of toxic wastes find it cheaper and easier to export such products than to comply with domestic regulations on their management and disposal. Beyond the risk of accidents during shipment, there exists a growing hazard in the practice of sending toxic wastes to financially hard-pressed developing nations that generally cannot safely handle the wastes being generated and stockpiled within their own borders.

3. It is not environmentally responsible to site an industrial operation which will produce hazardous waste in a location where there is no realistic means of disposal. In the extreme, this means that certain industrial subsectors may not be suitable in entire regions. The problem is often institutional and a site might be conditionally acceptable if the proposed development were accompanied by a governmental commitment and plan to provide for hazardous waste transport, treatment, and disposal.

4. The World Bank is committed to promoting sustainable development -- long-term development based on the effective management of natural and human resources. In view of the accelerating danger to the environment created by unregulated international toxic waste shipments and in light of its own fundamental concerns with environmental protection in developing nations, the World Bank's position on the handling, shipment, and disposal of toxic or hazardous waste is as follows:

   (a) Neither the oceans nor any developing country should be put at risk through dumping, transshipment, or disposal of toxic or hazardous wastes produced in one country and transported to another.

   (b) Ocean dumping must be strictly and completely prohibited. If employed at all, international shipment of toxic wastes must be conducted only with the prior, informed consent of competent government authorities; and after acceptable certification that the transport used meets international conventions and standards, that the shipment moves through safe facilities to appropriate and environmentally sound storage and disposal sites managed by experienced, responsible, and certified operators under adequate monitoring safeguards.

5. These are issues of international importance that require a strong, collaborative response. The Bank recognizes the stated concerns of the Organization for Economic Cooperation and Development (OECD) and Organization for African Unity (OAU), and such international efforts as those of the United Nations Environment Programme (UNEP) to establish the essential government-to-government arrangements for prohibiting or policing toxic and/or hazardous waste disposal. The World Bank is ready to cooperate with the international community and national governments to develop clear standards and codes of practice to ensure environmentally sound hazardous waste management.

6. Within the framework of the Bank's lending policies and operations, it will not finance any projects in any of its borrowing countries that involve the disposal of hazardous or toxic wastes from another country; nor will it finance the shipment of hazardous or toxic wastes to any developing country for disposal. While strongly endorsing national and international efforts to develop suitable environmental standards and codes of practice for the safe transport, transfer, storage and disposal of hazardous or toxic waste, the Bank will continue to support the efforts of borrowing countries to build or strengthen their own domestic facilities for effective waste mangement through recycling, recovery, reprocessing and safe disposal.

Asbestos in Bank-Financed Projects

7. The Bank recognizes that asbestos is a hazardous substance. The Bank increasingly prefers to avoid financing asbestos use. This preference will be updated as frequently as science progresses. Thus, at any mention of asbestos in Bank-assisted projects, the Task Manager needs to exercise special care.

8. The major risks to human health and occupational safety associated with asbestos products are well documented in the literature cited. The scientific evidence is mounting that long exposure is powerfully carcinogenic, but that the prospects of some workers may improve if exposure is subsequently reduced. The threshold below which exposure is risk-free is low, but unknown precisely.

9. Historically, the Bank has not directly financed the mining of asbestos. Given the present weight of scientific opinion, the Bank prefers not to finance the manufacture or use of asbestos-containing products. This preference is related to risk. On the one hand, the Bank is more likely to decline financing asbestos when used dry and exposed to the air where people may be at risk, such as in school roofing. On the other hand, this preference is less categorical where the product used is wet, stable, and underground as in sewerage pipes.

10. If there are to be any exceptions to this preference, the onus is on the project proponents and the Task Manager to justify waivers at or soon after IEPS, on a case by case basis. The criteria to be considered for such waivers will include the economics of alternatives, and the magnitude of the risks involved. The onus is on proponents to show the unavailability of alternatives, partly because acceptance of the use of lower risks asbestos-cement sewer pipes implies some incentive for asbestos manufacture and mining, which are higher risks and further outside project standards control.

11. Risk-assessment tied to best estimates of toxicity of different products and emissions may become available to devise a portfolio policy which could be applied to differential opportunity costs in different places. Market incentive policies may take precedence over quantitative controls for intermediate hazards. TMs should ascertain the status of such a portfolio policy when seeking waivers. The Environment and Industry Departments, the Science and Technology, and the Health advisors should be contacted for such waivers and for any amplification needed.

12. The Bank does not promote the premature demolition of existing asbestos-containing structures, particularly those containing dry and non-bound asbestos. The Bank is not only concerned with amphibole asbestos, but also with serpentine, chrysotile asbestos. Recent evidence (Anderson 1991) shows that white asbestos (chrysotile) also causes incurable cancer (mesothelioma) which may not appear for 30 to

50 years after exposure. The Bank is particularly concerned to avoid human exposure to the amphibole forms of crocidolite or "blue asbestos". Scientific evidence shows that this increases the health risks to demolishers and the public. While it may be possible to demolish such structures with acceptably low risks, this is unlikely to pertain widely in developing countries. Nor does the Bank encourage the premature replacement of existing asbestos-cement cold-water pipes because the risks from drinking such water are low compared with inhalations risks.

13. The Bank is willing to consider financing (a) alternatives to asbestos products, (b) monitoring and assessing asbestos health risks, (c) risk-reduction measures in existing asbestos-containing structures, including building maintenance, (d) safe disposal of asbestos-containing substances and (e) education, training, and safety measures.

PLANT SITING AND INDUSTRIAL ESTATE DEVELOPMENT

1. Sites for industrial plants, generating stations, municipal wastewater treatment plants, solid waste management systems, and similar facilities have historically been selected on the basis of economic and technical factors, including: favorable terrain, access to raw materials, energy sources, transportation and labor, location and size of markets or service areas, taxes and duties, and availability of utilities and other support services essential for successful plant operation. More recently, the siting of industry has evolved to include considerations of the natural and sociocultural environment and of acceptance by the communities that could be affected, either positively or negatively.

2. No longer is the goal of industrial growth sufficient as the sole justification for construction of a factory at a particular site. Increased knowledge of public health effects and experience with the degradation of air, water, and land that can occur in the absence of sound planning in industrial areas is one reason for the change. Community unwillingness to tolerate disturbance in forms such as noise, traffic, odors and physical presence of large facilities is another. At least equally important is public awareness of the hazards posed by many industrial operations, heightened by widely publicized disasters such as Bhopal and Chernobyl.

3. Industry, at the same time, has gained experience with the costs of routine pollution control and waste disposal operations, accident response and remedial clean-up activities. In countries where environmental standards are enforced, pollution control represents a known and substantial expense item. It then tends to be integrated into the economic decision-making of companies seeking new plant locations, in that sites with special environmental sensitivities requiring extraordinary measures to protect environmental quality become relatively less attractive economically. Conversely, pre-planned industrial estates with waste treatment and disposal systems and other necessary infrastructure offer distinct advantages.

4. As a consequence, it is becoming an increasingly common practice to require advance governmental approval of sites for industries that have potentially adverse impacts. India, for example, began in 1988 to require clearance from the cognizant State Pollution Control Board before a plant is sited, rather than simply before it begins to operate. The Indian Ministry of Industry has established a formal site selection procedure for highly polluting industries (Geethkrishnan 1989). Indonesia recently began requiring industries to conduct an environmental review of proposed new facilities. For those in which significant

impacts are identified, the firm must complete a full impact assessment before a license can be granted by the Ministry of Industry or the Investment Coordinating Board.

5. The trend is to integrate the determination of site suitability into the overall pollution control/environmental management process. In essence, many agencies responsible for pollution control, and even some of those which exist primarily to promote industrial development, have been pursuing the same objective as that articulated in the EA OD (para 4): "close integration of EA with... other aspects of project preparation" so that "environmental considerations are given due weight in project selection, siting, and design...."

The Practice of Industrial Site Selection

General Procedures

6. Although there are different methodologies employed for comparative siting, seven basic elements are common to all of them:

(a) A short list of potential sites (may include both preferred and alternative sites).

(b) Description of each site in terms of ecological and sociocultural sensitivities.

(c) Analysis of capacity to assimilate impacts at each site in terms of a common set of criteria for prevention of natural and sociocultural resource degradation.

(d) Elimination of sites with serious environmental limitations.

(e) For remaining sites, description of measures to avoid or mitigate impacts and comply with environmental standards, including consideration of technical and institutional feasibility, reliability and life-cycle cost.

(f) Consultation with affected communities.

(g) Ranking of alternatives and selection of proposed site.

Depending on the regulations of the country and the nature of the industry, the site selection process may be carried out in the context of an EA or as a more specific analysis under a licensing or permit application procedure.

7. Sites may be "preselected" as well, either as part of a planning and zoning process which narrows the range of alternatives to areas designated for industry, or under development policies which seek to localize industrial development in industrial estates. If planning and zoning and industrial estate siting are based on environmental criteria, there may be no need for additional siting analysis, or the studies required may be limited to particular issues, such as the need to pretreat a proposed plant's wastewater.

However, it is often true that only economic and engineering feasibility criteria are used as the basis for identifying areas for industrial development. There is then no guarantee that environmental objectives will be met. An environmental analysis of possible sites should be conducted.

8. In the case of a proposed expansion of production facilities at their present site, it is important to evaluate the site on the basis of the combined effects of the existing and new operations. Some unique feature of the new process may make the site undesirable, or the measures needed to manage the overall impact may be so costly that a new site is preferable. The same concept applies to location of a new plant in an already industrialized area. The incremental increase in cumulative air emissions, for example, may make the site unacceptable for the proposed facility.

Critical Industrial Subsectors

9. Site selection is not a critical issue for all industrial subsectors. Either because of smaller size or type of operation, or both, there are plants which have little potential for adverse impact on natural or sociocultural surroundings. Indonesia accounts for this by having a two-level environmental review procedure. All enterprises applying for licenses must complete a preliminary assessment. Those activities for which potentially severe impacts are identified will require that a full impact assessment be conducted. India has taken an alternative approach. The Ministry of Industry publishes a list of twenty subsectors for which formal environmental clearance by state government is required for siting. Some examples are: primary metallurgical industries, pulp and paper, paints, leather tanning, storage batteries, synthetic rubber, cement, and electroplating. Large installations would be obliged to conduct full environmental impact assessments in accordance with the regulations of the Indian Ministry of Environment and Forests (Geethkrishnan 1989).

Siting Criteria

10. Sites should be compared and selected on the basis of a comprehensive set of siting criteria. Sometimes industrial siting criteria may already exist in the form of government regulation or guidelines. Where they do not exist in this form, they can be derived for the project from various sources. Siting criteria may be implicit in planning and zoning, as the basis for determining suitability for industrial land uses. Laws or regulations for protection of certain sensitive areas or resources act as restrictions on and should be incorporated in the criteria used in site selection. There are criteria considered to represent good practices for particular industries. Finally, there are the general principles of environmentally-sensitive land use planning.

11. Continuing with the Indian example, the Ministry of Environment and Forests has recommended guidelines addressing (a) areas to be avoided and (b) environmental requirements for industrial sites. Examples are given below (Geethkrishnan 1989):

(a) An industrial site shall be at least the following distances from the features listed:

- 25 km from ecologically or otherwise sensitive areas (examples include religious and historic places and archaeological monuments, scenic areas, beach resorts, coastal areas and estuaries which are important breeding grounds, national parks and sanctuaries, natural lakes and swamps, and tribal settlements)

- 0.5 km from high tide line in coastal areas
- 0.5 km from natural or modified flood plain boundary
- 25 km from projected growth boundary of major settlements (population of 3 million or larger)

(b) The following are examples of environmental requirements associated with industrial use of particular sites:

- no conversion of forest land to non-forest activity to sustain the industry
- no conversion of prime agricultural land to industrial use
- sufficient space on-site to provide for storage of solid waste and appropriate treatment and reuse of wastewater
- provision for a 0.5-km wide "greenbelt" around the site perimeter
- adaptability of the proposed facilities to the landscape, so that scenic features are not altered by the development

12. The Indian guidelines may not be applicable in every country, but they illustrate the kinds of considerations governments are requiring in industrial siting. Examples of other factors that might be placed in a list of characteristics precluding selection of a particular site for use by industry with high potential for pollution include (depending on the nature of the industry):

(a) Recharge area for aquifer of present or possible water supply use, or catchment area of public water supply reservoir.

(b) Receiving waters unable to assimilate wastewater without water quality degradation despite appropriate treatment.

(c) Airshed prone to episodes of unhealthful air quality.

(d) Habitat of endangered species.

(e) Proximity of site (or access roads) to incompatible land uses -- e.g., health care institutions, schools, residential areas.

(f) No local or regional capability for disposal of hazardous waste (if industry produces any).

13. There are other factors which ordinarily do not exclude a site from consideration, but which are potential areas of impact and should be taken into account in ranking alternative sites.

- number of residents that would be displaced;
- number of properties that would be affected or expropriated;
- distance to nearest non-industrial land use; and
- compatibility of wastewater with local collection and treatment system, if any.

Relationship to Environmental Assessment

14. Facility siting is one of the areas in which EA can be most effective, but only if the assessment process begins before siting options are foreclosed. Complex industrial development projects and similar facilities with the potential for significant environmental impact cannot be handled with the simple application of siting criteria. Such projects need a full EA. The EA should be initiated well before the siting decision has been made, so that real alternatives can be considered. Identifying the potential impacts associated with each site and comparing sites on that basis causes environmental issues to come to light early and permits project planners and designers to take maximum advantage of all possible ways to avoid impacts. For those impacts that cannot be avoided and are accepted as part of the costs of the development, the opportunity to select an alternative site may lead to a project in which the efficiency of measures to mitigate impacts is higher and the costs of the measures are lower than would otherwise be the case. A timely EA also prevents the disruption, delay, and extra expense involved when a site must be changed because of environmental or public acceptance issues that come to light during final design.

Special Considerations in Industrial Plant Siting

Assimilative Capacity of the Environment

15. The most obvious example of failure to consider assimilative capacity is the siting of a manufacturing plant (e.g, a pulp mill) where flow in the receiving stream is seasonally or perennially less than the quantity of effluent. Unless the effluent can be treated (or cooled, if it is cooling water) to be of equal or better quality than the receiving water, disturbance of the aquatic ecosystem is inevitable. Such treatment, if technically achievable at all, is likely to be extremely costly. An alternative site, with receiving waters capable of accepting properly treated effluent without significant degradation, could lead to lower cost over the life of the facility. The same may be true of sites where water supply is limited or where meteorological conditions (e.g., frequent atmospheric inversions) would necessitate unusually stringent waste treatment practices.

16. Another aspect is the ability of the environment to assimilate the results of non-routine operations, such as process upsets, failure of pollution control systems, and accidental releases. Proximity to sensitive natural areas or human settlements may necessitate extraordinary measures to prevent or respond to such events.

Area of Influence

17. Depending on the type of facility and the medium being considered (air, water, plant, animal or human communities), the area that might be influenced by a project can extend well beyond the site and its immediate environs. Effects on water availability at the point of withdrawal and on receiving water quality for some distance downstream of the point of discharge are factors that must be considered in site selection. The characteristics of the natural resources and land uses in the airshed for long distances downwind are relevant and so are environmental impacts along transportation corridors. If the project would result in ancillary developments that would differ depending on site selection (e.g., asphalt plants

at quarry sites, new rail or roadways, new port facilities or pipelines, workers dwellings, resettlement sites), their water catchments and airsheds should be considered in the siting decision.

Capacity for Emergency Response

18. It is similarly irresponsible to locate an industrial plant which poses a significant risk to neighboring communities or sensitive natural systems in surroundings where an emergency cannot be managed in such a way that damage or disaster can be averted. If it is not possible to develop a response plan which can reasonably be expected to be effective (including provisions for emergency evacuation, if warranted by the type of installation), another site should be selected. The absence of institutions for communication and accident response makes hazard management impossible. Unsafe roads or railways and unsafe trucks or trains lead to unacceptable risk, if they are used to transport hazardous substances through residential areas. Lack of a buffer zone between hazardous material storage or processing facilities and communities or sensitive natural systems (fish breeding areas, for example) creates a situation in which neither warning nor containment can be timely enough to prevent injury.

19. Some of these limitations can be overcome by adding hazard management components to a project. Local government's response capacity can be strengthened by providing equipment and training. Transportation facilities can be improved, or alternate routes to the site can be developed. However, some dimensions of the emergency response problem can only be resolved through sound site selection. Further information on this topic is provided in the section on "Industrial Hazard Management."

Induced Development

20. Employment opportunities are magnets to immigration of workers and thus to the growth of local communities. Especially where industrial development is newly occurring, the community may experience induced land development and may be ill-prepared to manage its impacts. They range from overloading of municipal infrastructure and services to cultural conflicts between long-time residents and immigrant workers. Particular care is needed to prevent unplanned settlements just outside the factory gates. Institutional strengthening of local government and involvement of local communities in project preparation can be effective ways of minimizing these adverse impacts. The section on "Induced Development" provides more information on this subject.

Community Involvement in Industrial Plant Siting

21. Community participation in siting decisions by private industry is far less common than in public investment projects but is still required, if Bank financing is involved. Businesses which have involved local residents early in decisions that may affect them, even on controversial projects, have more often than not found the experience to be worthwhile. Conducted well, it leads to better mutual understanding and can be the basis of productive community relations instead of protest. The EA OD states the Bank's expectation that the view of affected groups will be taken into account in project design and implementation. The siting process is an excellent place to begin. (See Chapter 7 for guidance on community involvement.)

ELECTRIC POWER TRANSMISSION SYSTEMS

1. Electric power transmission systems include the transmission line, its right-of-way (ROW), switchyards, substations, and access or maintenance roads. The principal structures of the transmission line include the line itself, conductors, towers, and supports (e.g., guy wires).

2. The voltage and capacity of the transmission line affects the sizes required for these principal structures. For example, the tower structure will vary directly with the required voltage and capacity of the line. Towers can be single wood pole structures for small voltage transmission lines up to 46 kilovolts (kV). H-frame wood pole structures are used for line ranging from 69 to 231 kV. Self-supporting single-circuit steel structures are used for 161 kV or larger lines. Up to 1,000 kV transmission lines are possible.

3. Transmission lines can range from several kilometers to hundreds of kilometers in length. The ROW in which the transmission line is constructed can range in width from 20 meters to 500 meters or greater depending upon the size of the line and the number of transmission lines located within the ROW. Transmission lines are primarily overland systems and can be constructed to span or cross wetlands, streams, rivers, and nearshore areas of lakes, bays etc. Underground transmission lines are technically feasible but are very costly.

Potential Environmental Impacts

4. Electric power transmission lines are linear facilities that will affect natural and sociocultural resources. The effects of short transmission lines can be localized; however, long transmission lines can have regional effects. In general, the environmental impacts to natural, social, and cultural resources increase with increasing line length. As linear facilities, the impacts of transmission lines occur primarily within or in the immediate vicinity of the ROW. The magnitude and significance of the impacts increase as the voltage of the line increases, requiring larger supporting structures and ROWs. Operational impacts also increase. For example, electromagnetic field (EMF) effects are significantly greater for 1,000-kV lines than for 69-kV lines.

5. Negative environmental impacts of transmission lines are caused by construction, operation and maintenance of transmission lines. Clearing of vegetation from sites and ROWs and construction of access roads, tower pads, and substations are the primary sources of construction-related impacts (see Table 10.2 at the end of this section for a summary of all potential impacts). Operation and maintenance of the transmission line involves chemical or mechanical control of vegetation in the ROW and occasional line repair and maintenance. These, plus the physical presence of the line itself, can be a source of environmental impact.

6. On the positive side, power line ROWs, when properly managed, can be beneficial to wildlife. Cleared areas can provide feeding and nesting sites for birds and mammals. The "edge" effect is well-documented in biological literature; it describes the increased habitat diversity resulting at the contact between the ROW and the existing vegetation. Power lines and structures can serve as nesting sites and perches for many birds, especially raptors.

Special Issues

Effects on Land Use

7. Electric power transmission lines have the greatest impact on land resources. A dedicated electric power transmission line ROW is required. Grazing and other agricultural uses are usually not precluded in ROWs, but other uses are generally not compatible. Although ROWs are generally not very wide, they can interfere with or fragment existing land uses along the ROW. Long transmission lines will affect more areas and result in more significant impacts.

8. Transmission lines can open up more remote lands to human activities such as settlement, agriculture, hunting, recreation, etc. Construction of the ROW can result in the loss and fragmentation of habitat and vegetation along the ROW. These effects can be significant if natural areas, such as wetlands or wildlands are affected, or if the newly-accessible lands are the home of indigenous peoples.

Clearing and Control of Vegetation in Rights-of-Way

9. A variety of techniques exist for clearing vegetation from the ROW and controlling the amount and type of new plant growth. From an environmental point of view, selective clearing using mechanical means or herbicides is preferable and should be evaluated in project EAs. Broadcast aerial spraying of herbicides should be avoided because it affords no selectivity, releases unnecessarily large amounts of chemicals into the environment, and because it is an imprecise application technique, may result in contamination of surface waters and terrestrial food chains, as well as elimination of desirable species and direct poisoning of wildlife.

Health and Safety Hazards

10. Placement of low-slung lines or lines near human activity (e.g., highways, buildings) increases the risk for electrocutions. Technical guidelines for design ordinarily minimize this hazard. Towers and transmission lines can disrupt airplane flight paths in and near airports and endanger low-flying airplanes, especially those used in agricultural management activities.

11. Electric power transmission lines create electromagnetic fields (EMFs). The strengths of both electric and magnetic fields decrease with distance (e.g., meters) from transmission lines. The scientific community has not reached consensus on specific biological responses to EMF, but the evidence suggests that health hazards may exist. Several states in the United States have promulgated rules regulating EMFs associated with high-voltage transmission lines.

Induced Development

12. Depending on their location, transmission lines may induce development in or bordering on ROWs or in lands made more accessible. In locales where the supply or housing is limited, cleared ROWs are often attractive sites for unpermitted housing, which in turn gives rise to other environmental impacts and overburdens local infrastructure and public services.

Project Alternatives

13. The environmental assessment should include an analysis of reasonable alternatives to meet the ultimate project objective of the distribution of electricity to load centers. The analysis may lead to alternatives which are more sound from an environmental, sociocultural, and economic point of view than the originally proposed project. A number of alternatives need to be considered, including:

- taking no action to meet the needed capacity
- alternative voltages
- DC transmission lines (permitting narrower ROWs)
- alternative sources of electricity
- construction of smaller power facilities closer to the loads as an alternative to bulk power transmission
- energy and load management plans to reduce need for additional power
- upgrading of existing facilities
- alternative routes and substation sites
- underground transmission lines
- alternative methods of construction including costs and reliability
- alternative transmission tower design and materials (e.g., wooden poles, steel or aluminum structures, etc.)
- alternative maintenance techniques and road designs

14. One of the most important considerations is an evaluation of alternative routes and substations sites. Many of the environmental impacts resulting from electric power transmission lines can be avoided or minimized by careful ROW and substation site selection.

Management and Training

15. The most critical environmental decision associated with electrical power transmission line construction and operation is the route selection. Environmental scientists need to work with the transmission line engineers in route selection and development of mitigative measures. Depending on the education and experience of the staff, training in the environmental management of electrical power transmission lines may be warranted. The major environmental specialties related to the environmental management of electrical power transmission lines are ecological impact and social impact assessments. Environmental training and management may be warranted for ROW maintenance techniques, including the proper use of chemical and mechanical clearing methods.

16. The training should be done as part of the environmental assessment phase of the project and with assistance from the environmental consultant. If at all possible, the project sponsor's environmental staff should be involved in the environmental assessment study. This will ensure an understanding of the environmental aspects of the project. In particular, staff workers must have an understanding of the rationale for the recommended mitigation and monitoring that they may be implementing.

17. Local, regional, and national environmental agencies involved in the review and approval of the project may also need training to monitor and enforce compliance during the construction and operation of the project.

Monitoring

18. The monitoring requirements for transmission lines will be dependent on the type of environmental resources involved and the degree to which they are affected. Monitoring construction activities may be required to assure that negative land use and/or ecological impacts are avoided and proper mitigation measures are employed. Monitoring of these impacts will be short-term (e.g., weeks) and occur along the line as it is constructed. Monitoring may be especially critical at crossings of major water bodies or wetlands, near wildlands and cultural properties. The actual monitoring will be based on visual inspections of the materials being used, the construction practices, and mitigation measures. Monitoring of ROW maintenance activities is also to be required to assure proper vegetation control methods, to prevent invasion of exotic species, and to support decisions which take advantage of possible benefits to wildlife.

Table 10.2. Electric Power Transmission Systems

Potential Negative Impacts	Mitigating Measures

Direct

1. Vegetation damage, habitat loss, and invasion by exotic species along the ROW and access roads and around substation sites.
   - Utilize appropriate clearing techniques, (e.g., hand clearing versus mechanized clearing).
   - Maintain native ground cover beneath lines.
   - Replant disturbed sites.
   - Manage ROWs to maximize wildlife benefits.

2. Habitat fragmentation or disturbance.
   - Select ROW to avoid important natural areas such as wildlands and sensitive habitats.
   - Maintain habitat (i.e., native vegetation) beneath lines.
   - Make provisions to avoid interfering with natural fire regimes.

3. Increased access to wildlands.
   - Select ROW to avoid sensitive wildlands.
   - Develop protection and management plans for these areas.
   - Use discontinuous maintenance roads.

4. Runoff and sedimentation from grading for access roads, tower pads, and substation facilities, and alteration of hydrological patterns due to maintenance roads.
   - Select ROW to avoid impacts to water bodies, floodplains, and wetlands.
   - Install sediment traps or screens to control runoff and sedimentation.
   - Minimize use of fill dirt.
   - Use ample culverts.
   - Design drainage ditches to avoid affecting nearby lands.

Table 10.2. Electric Power Transmission Systems (continued)

Potential Negative Impacts	Mitigating Measures
Direct (continued)	
5. Loss of land use and population relocation due to placement of towers and substations.	5. • Select ROW to avoid important social, agricultural, and cultural resources. • Utilize alternative tower designs to reduce ROW width requirements and minimize land use impacts. • Adjust the length of the span to avoid site-specific tower pad impacts. • Manage resettlement in accordance with Bank procedures.
6. Chemical contamination from chemical maintenance techniques.	6. • Utilize mechanical clearing techniques, grazing and/or selective chemical applications. • Select herbicides with minimal undesired effects. • Do not apply herbicides with broadcast aerial spraying. • Maintain naturally low-growing vegetation along ROW.
7. Avian hazards from transmission lines and towers.	7. • Select ROW to avoid important bird habitats and flight routes. • Install towers and lines to minimize risk for avian hazards. • Install deflectors on lines in areas with potential for bird collisions.
8. Aircraft hazards from transmission lines and towers.	8. • Select ROW to avoid airport flight paths. • Install markers to minimize risk of low-flying aircraft.
9. Induced effects from electromagnetic fields.	9. • Select ROW to avoid areas of human activity.

Table 10.2. Electric Power Transmission Systems (continued)

Potential Negative Impacts	Mitigating Measures
Direct (continued)	
10. Impaired cultural or aesthetic resources because of visual impacts.	10. • Select ROW to avoid sensitive areas, including tourist sites and vistas. • Construct visual buffers. • Select appropriate support structure design, materials, and finishes. • Use lower voltage, DC system, or underground cable to reduce or eliminate visual impacts of lines, structures, and ROWs.
Indirect	
1. Induced secondary development during construction in the surrounding area.	1. • Provide comprehensive plans for handling induced development. • Construct facilities to reduce demand. • Provide technical assistance in land use planning and control to local governments.
2. Increased access to wildlands.	2. • Route ROW away from wildlands. • Provide access control.

OIL AND GAS PIPELINES

1. Oil and gas pipeline projects include the construction and operation of offshore, nearshore and/or overland pipelines. Pipelines can range in size up to 2 meters in diameter. They can range in length from several kilometers to hundreds of kilometers. Overland and nearshore pipelines are generally buried. Offshore pipelines are generally located on the seafloor in waters as deep as 350 to 450 meters, but subsea pipelines have been laid below 1,500 meters in special cases.

2. The major facilities associated with oil or gas pipelines include the pipeline itself, access or maintenance roads, the receiving, dispatch and control station, and the compressor station or pump stations. Because of internal friction and changes in elevation encountered along the line, booster stations are required for long-distance crude oil and product pipelines at regular intervals (e.g., approximately 70 kilometers [km]). Compression stations are installed at appropriate intervals along gas transmission lines to maintain pipeline pressures. The pipeline may transport unrefined oil or gas from a wellhead to transfer or processing facilities. Refined oil and gas may be transported by pipeline to an end user, such as a petrochemical plant or power plant.

Potential Environmental Impacts

3. Pipeline installation in upland areas involves surveying, right-of-way (ROW) clearing, ditching, pipe stringing, bending, welding, wrapping, coating and installing cathodic protection for corrosion control, placement in ditch (for buried pipelines), backfilling and cleanup. The same general activities occur in wetland areas, but dredging and spoil disposal are necessary for placement of the pipeline. In completely water-logged soils and open-water areas, pipeline-laying barges are used for dredging, pipeline fabrication and placement.

4. Installation of pipelines in offshore areas involves laying the pipeline on the bottom. The pipeline may be anchored with concrete blocks or concrete casing. If the pipeline is to be buried, then digging the trench is required. Pipeline is laid by a laying barge. Trenching is accomplished by underwater trenching machines. Most often the natural processes of current and wave action are relied on for the burial of pipelines in offshore areas, but artificial burial can be accomplished. In nearshore/landfall areas, buried pipelines are required.

5. Proper pipeline operation emphasizes maintenance and checking of equipment. Periodic ground or aerial inspection along the pipeline route is required to detect leaks. Devices used to scour or clean paraffin and scale from the inside of oil pipelines (referred to as scrapers, balls, and "pigs") or to separate materials pumped through the pipeline or to remove liquids and condensate (in gas pipelines) can result in wastes that must be disposed of. Pipeline life depends on the rate of corrosion and inside wear of the pipe. Corrosion protection is a necessity in most soils, especially in wet or saline areas. Leaks or ruptures of oil and gas pipelines can have significant impacts beyond the immediate vicinity of the pipeline.

Positive Impacts

6. In some cases, oil and gas pipelines may be viewed as contributing to environmental quality by making cleaner fuels more available (e.g., low sulfur gas versus high sulfur coal) for energy production and/or industrial purposes. In offshore areas, unburied pipelines may create habitat for marine organisms attracted to the new "artificial reef".

Negative Impacts

7. Offshore, nearshore, and upland oil and gas pipelines have different environmental impacts according to type, as amplified in the paragraphs below. (See also Table 10.3 at the end of this section for other examples.) The magnitude of their impacts depends on the type and size of the pipeline installed; the significance depends on the degree to which natural and social resources are affected.

8. **Direct Impacts: Offshore Pipelines**

    (a) Installation of pipelines in offshore and nearshore areas may result in the loss of benthic and bottom-feeding organisms from trenching and/or turbidity associated with pipeline laying. The significance of these impacts will depend on the type of aquatic resources affected and the extent these resources are affected.

    (b) The construction of the pipeline can result in the temporary resuspension of bottom sediments. The redisposition of sediment may alter aquatic habitat characteristics and lead to changes in species composition. The significance of these effects will depend on the type and importance of aquatic organisms affected. For example, the significance of habitat alteration of seagrass beds or coral reefs, considered important feeding and breeding habitats for fish and other animals, may be greater than the alteration of the deep offshore benthic habitat.

    (c) If pipeline trenching occurs in nearshore and offshore areas where toxic chemicals have accumulated in the sediments (e.g., harbors near industrial outfalls of toxic chemicals such as mercury and polychlorinated biphenyl [PCBs]), the laying of the pipeline can result in a resuspension of these toxic sediments and temporarily lower water quality immediately above the pipeline. Bioaccumulation of these toxic chemicals may occur in aquatic organisms (e.g., fish and shellfish).

    (d) In nearshore and offshore areas used for bottom fishing, pipelines can interfere with bottom trawling, resulting in loss or damage to fishing equipment as well as accidental ruptures to pipelines. Anchor dragging can also result in pipeline damage and oil spills.

9. **Direct: Upland Pipelines**

    (a) Installations of pipelines can lead to erosion in the vicinity of the pipeline. In hilly areas, this can lead to instability in the soils and landslides. Runoff and sedimentation can lower water quality in rivers and streams during construction.

(b) Installation of pipelines and maintenance roads can lead to alteration of drainage patterns, including blocking water flow and raising the water table on the upslope side of the pipeline, and can lead to the killing and reduction of vegetation, such as trees. If a pipeline cuts through a large forested area, this impact could be significant. Water supply to wetlands can be altered.

(c) Creation of ROWs can lead to the invasion of exotic plants which may out-compete native vegetation. If uncontrolled, this can have a significant impact over time. In addition, pipeline installation can result in habitat fragmentation of natural areas (e.g., wildlands), resulting in the loss of species and lowering of biodiversity.

(d) In developed areas, oil and gas pipelines can result in the loss of land use and displacement of inhabitants due to the placement of pipeline and substations. Some types of agricultural activities may only be affected in the short-term during construction.

(e) Above-ground pipelines can create barriers for humans and migratory wildlife. This could be significant depending upon the length and location of the pipeline.

(f) Archeological sites are vulnerable to damage or loss during pipeline construction.

(g) Pipeline construction can cause temporary interruption of traffic. This could be significant in developed areas, if the pipeline crosses major transportation routes.

(h) Ruptures and leaks, as well as wastes generated at the pump and transfer stations, can result in potential contamination of soils, surface water and groundwater. The significance of this contamination is dependent on the type and size of the leak, on the type and volume of wastes generated, and on the degree to which the natural resource is affected. Ruptures of oil pipelines crossing rivers and other water bodies or wetlands can result in significant environmental damage.

(i) Gas pipeline leakage or rupture can cause explosions or fires. In developed areas, such accidents pose significant human health risk.

10. **Indirect Impacts**

(a) Upland pipelines can result in inducing secondary development (e.g., squatters) within the pipeline ROW. This unplanned development can place strain on the existing infrastructure for an affected area.

(b) Upland pipelines can allow access to otherwise inaccessible natural areas (e.g., wildlands). This can result in degradation and exploitation of these areas.

Special Issues

Natural Resources

11. Marine and estuarine water resources are affected by offshore and nearshore oil and gas pipelines. Freshwater resources can be affected by upland pipelines. Depending upon the ROW location, the construction of the pipeline in and near streams, rivers, lakes, and estuaries could result in significant water quality impacts from sedimentation and runoff. In addition, the flood storage functions of these systems can be altered by changing surface water drainage and by construction of facilities within these water bodies.

12. Construction of undersea pipelines may impact significant coastal and marine resource (e.g., coral reefs, sea grass beds, etc.), and affect fishing activities. Ruptures of the oil pipeline or accidental spills of oil at terminals would significantly affect water quality in streams, rivers, lakes, estuaries, and other water bodies along the pipeline ROW. Possible groundwater contamination could occur from these spills, depending on the type and extent of the spill and the hydrogeological features of the area.

13. Long pipelines can open up less accessible natural areas, such as wildlands, to human activity (agriculture, hunting, recreation, etc.). Depending on the tolerance of the ecological resources in these areas and the sociocultural characteristics of the population, these activities may have adverse impact.

Pipeline Safety

14. The transportation of natural gas by pipeline involves some degree of risk to the public in the event of an accident and subsequent release of gas. The greatest hazard is a fire or explosion following a major pipeline rupture.

15. The primary cause of pipeline accidents is outside forces, implicated in more than half of all service incidents. Other causes include corrosion and material and construction defects. Accidents result from careless operation of mechanical equipment (bulldozers and backhoes), earth movements due to soil settlement, washouts, landslides or earthquakes, weather effects (winds, storms, thermal strains) and deliberate damage. Some countries have national safety standards for the construction and operation of gas pipelines. The World Bank has environmental guidelines for the construction and operation of oil pipelines.

Other Special Issues

16. Depending on the location, oil and gas pipelines may impact cultural properties, land settlement, tribal peoples, biological diversity, tropical forests, watersheds, and wildlands. World Bank policies and guidelines have been developed for these impacts (see relevant discussions in Chapters 2 and 3).

Project Alternatives

17. The environmental assessment for oil or gas pipeline should include an analysis of reasonable alternatives to meet the ultimate project objective. The alternative analysis may lead to designs that are more sound from an environmental, social, and economic point of view than the originally proposed project. The following alternatives should be considered:

- the "no action" alternative (i.e., examine the feasibility of taking no action to meet the needed fuel capacity)
- alternative means of delivering the oil or gas (e.g., tankers)
- upgrading existing facilities
- alternative routes and substation sites
- alternative methods of pipeline construction including costs and reliability
- alternative pipeline design and materials (e.g., buried versus elevated pipelines)

18. The appropriateness and inappropriateness of these alternatives should be addressed in relation to environmental and economic factors. Because oil and gas pipelines are linear facilities, one of the most important alternatives available is in choice of routes. Many of the environmental impacts associated with oil and gas pipelines can be avoided or minimized by careful route selection.

Management and Training

19. As discussed in the preceding paragraph, one of the most critical environmental decisions associated with oil and gas pipeline construction and operation is the route selection. Environmental scientists should work with the pipeline engineers to select routes and develop mitigative measures.

20. Depending on the education and experience of the staff, training in the environmental management of oil and gas pipelines may be warranted. In particular, staff workers must have an understanding of the rationale for the recommended mitigation measures and monitoring program that they may be implementing. Local, regional, and national environmental agencies involved in the review and approval of the project may also need training to monitor and enforce compliance with environmental management requirements of the project.

21. Safety training and education may be necessary, including evacuation procedures and containment plans for oil spills and gas ruptures. An emergency response plan may be required in areas where there is risk to the public from accidents.

Monitoring

22. The monitoring requirements of oil or gas pipelines will be dependent on the type of environmental resources and the degree to which they are affected. Monitoring of construction activities

will be required to ensure compliance with good practices and any special requirements to avoid or mitigate adverse impacts and to detect any impacts which occur so that corrective action can be initiated. Material storage and equipment repair yards and construction worker camps should be included. The actual monitoring may range from visual inspection of the mitigation system (e.g., sediment traps) to more extensive water quality monitoring during the pipeline construction across or near a water body. If pipeline construction involves the potential for resuspension of toxic substances, an extensive chemical and biological monitoring program may be required.

23. Monitoring should occur before, during, and for some period after the pipeline is laid or buried. The objective of this monitoring program will be determined by the extent and duration of the recontamination of the water body. Monitoring of the operation of the oil and gas pipelines will be required to assure proper mechanical functions or to identify structural conditions resulting in leaks or ruptures.

Table 10.3. Oil and Gas Pipelines

Potential Negative Impacts	Mitigating Measures
Direct	
1. Resuspension of toxic sediments from construction of offshore pipelines.	1. • Select alternate location for laying pipeline. • Use alternative pipeline construction techniques to minimize resuspension of sediments (e.g., laying pipeline versus burying pipeline). • Lay pipeline at a period of minimal circulation.
2. Interference with fishing activities from offshore and nearshore pipelines.	2. • Select pipeline route away from known fishing areas. • Mark and map location of offshore pipelines. • Bury pipeline that must be located in critical fishing areas.
3. Habitat and organism loss along offshore and upland pipeline ROWs and pumping and compressor station sites, and increased access to wildlands.	3. • Select ROW to avoid important natural resource areas. • Utilize appropriate clearing techniques (e.g., hand clearing versus mechanized clearing) along upland ROWs to maintain native vegetation near pipeline. • Replant disturbed sites. • Use alternative construction techniques (see No. 1).
4. Erosion, runoff, and sedimentation from construction of pipeline, grading for access roads and substation facilities.	4. • Select ROW to avoid impacts to water bodies and hilly areas. • Install sediment traps or screens to control runoff and sedimentation. • Use alternative pipeline laying techniques that minimizes impacts. • Stabilize soils mechanically or chemically to reduce erosion potential.
5. Alteration of hydrological patterns.	5. • Select ROW to avoid wetlands and flood plains. • Minimize use of fill. • Design drainage to avoid affecting nearby lands.

Table 10.3. Oil and Gas Pipelines (continued)

Potential Negative Impacts	Mitigating Measures
Direct (continued)	
6. Evasion of exotic species and habitat fragmentation.	6. • Select corridor and ROW to avoid important wildlands and sensitive habitats. • Maintain native ground cover (vegetation) above pipeline. • Make provisions to avoid interfering with natural fire regimes.
7. Loss of land use due to placement of upland pipeline and substations.	7. • Select ROW to avoid important social (including agricultural) and cultural land uses. • Design construction to reduce ROW requirements. • Minimize offsite land use impacts during construction. • For buried pipelines, restore disturbed land along ROW.
8. Creation of barriers for human and wildlife movement.	8. • Select ROW to avoid travel routes and wildlife corridors. • Elevate or bury pipeline to allow for movement.
9. Increased traffic due to construction.	9. • Phase construction activities to control traffic. • Construct alternative traffic routes.
10. Chemical contamination from wastes and accidental oil spills.	10. • Develop waste and spill prevention and cleanup plans. • Utilize spill containment techniques. • Clean up and restore affected areas.
11. Hazards from gas pipeline leakage or rupture.	11. • Clearly mark locations of buried pipelines in high-use areas. • Develop emergency evacuation plans and procedures. • Monitor for leaks. • Install alarms to notify the public of accidents.

Table 10.3. Oil and Gas Pipelines (continued)

Potential Negative Impacts	Mitigating Measures
Indirect	
1. Induced secondary development during construction in the surrounding area.	1. • Develop comprehensive plan for location of secondary development. • Construct facilities and provide financial support to existing infrastructure.
2. Increased access to wildlands.	2. • Develop protection and management plans for these areas. • Construction barriers (e.g., fences) to prohibit access to sensitive wildlands.

OIL AND GAS DEVELOPMENT--OFFSHORE

1. This category includes exploration, development and production of offshore oil and gas resources. Major development phases include the initial geophysical surveys of broad regions to identify exploration targets, drilling wells from ships or temporary platforms to test likely targets, spaced development drilling from fixed production platforms, and construction of the transportation and processing infrastructure. Production units may include various types of platforms with multiple production and re-injection wells, storage tanks, separators, and ancillary support facilities. Transport is usually by pipeline, occasionally by barge or tanker, to shore-based refineries and/or gas processing facilities.

Potential Environmental Impacts

2. Exploration typically consists of geophysical surveys from aerial overflights and/or ship on line or grid traverses over broad areas, bottom sampling by various methods, seismic surveys with explosives or various concussion devices, and test drilling for geological data. This is followed by drilling selected structures from drill-ships or temporary platforms, off-set wells to delineate any oil and gas discoveries, and extensive well and production tests to determine resource parameters. Initial wells are stubbed off and capped pending production. A production complex may include one or more production platforms with both production and injection wells, primary processing and storage facilities, associated drilling platforms, submersible units, and pipelines for gathering and transportation to shore.

3. The production and drilling platforms are self-contained facilities with heliopads, living accommodations for work crews, power supplies, storage tanks, etc. Production requires an extensive shore-based support system for permanent housing of the work force, supplies, waste disposal and refining. Platforms and drill ships are supplied by both ship and air transport. Initial production is often transported to shore by tanker or barge. This practice may continue for small fields where a pipeline is not economical. (For guidance, see Table 10.4: "Checklist for Oil and Gas Development--Offshore" projects.)

4. Effluent discharges include treated sanitary and domestic wastes, treated drilling muds and cuttings, produced waters, and onshore nonpoint and point sources. Offshore, air emissions result from diesel-powered generators and pumps, blowouts with fire or release of sour gas, and operational emissions from transfer. Onshore, air emissions result from operation of oil refineries, gas processing plants and vessel offloading. Noise, normal to the operation of a large industrial complex, is continual at both offshore and onshore facilities. (See Table 10.5 at the end of this section for other examples of potentially negative environmental impacts from Oil and Gas Development--Offshore projects.)

5. Nonroutine catastrophic events that may occur include blowouts with fire or release of sour gas (hydrogen sulfide), platform collapse, pipeline break, and tanker collision.

Table 10.4. Checklist--Offshore Oil and Gas Development

1. **Production**

   - Field: Size, depth, area, structure, oil/gas/water ratios, oil type, gas type(s), pressures.

   - Operations: Site preparation, well spacing, start-up period (production rate, field life, sanitary wastes), pollution control, monitoring, oil spill and hydrogen sulfide contingency plans.

   - Air Emissions: Emission quantity and, where applicable, composition: venting, flaring, equipment emissions, evaporation from oil spills and leakage.

   - Waste Discharge: Projected quantity and composition; treatment/disposal method (production water, sanitary wastes, drilling muds and cuttings, oil spill and leakage).

   - Landuse: Area of field, port facilities, pipelines.

   - Equipment: Type and number of drilling and production platforms and ancillary units, transport of supplies and workers.

   - Supplies: Drilling muds, pipe, chemicals, water, fuel.

   - Staffing: Number and skills, source, housing plans.

2. **Environmental Resources**

   - Geology: Stratigraphy, structure, fracture patterns, aquifers (depth, thickness and quality, esp. if near shore), bottom character, geologic hazards, seismic history.

   - Oceanography: Water depths, temperature, mixing, tides and currents, bottom sediments, organic material, particulates, nutrients, salinity, contaminants.

   - Biological: Coastal habitats (coastal barriers, wetlands, bays, lagoons, estuaries, marshes, mangrove swamps, seagrasses); offshore habitats (shelf, banks, slope, deep sea, reefs); substrate, biota, communities, resident and casual populations, rare or significant species, significant habitat.

   - Climate: Precipitation patterns (amount, frequency, type), air quality, wind and storm patterns (direction, speed, frequency), temperature, climatic zone.

3. **Socio-Economic Factors**

   - Nearby communities: Location, access, population (number, demographic and social characteristics); economy (employment rate, income distribution, tax base); services (types, capacity, adequacy) and housing; concern is the ability to (a) provide workforce, (b) service new development and (c) absorb and adjust to growth (worker/family in-migration).

   - Land Use: Intensive and casual, full time and seasonal, actual and projected, specially designated areas (marine sanctuaries, coral reefs, recreational beaches or seashores, parks, refuges, reservations, wilderness), man-made features.

   - Cultural: Historic sites, archaeologic sites, native religious or harvest sites, ship wrecks.

4. **Regulatory Framework**

   - Applicable environmental laws, regulations, policies, standards, and requirements; monitoring and enforcement: air, water, waste, noise, reclamation, land use controls and approvals, cultural and historic resource protection.

   - Designation and protection of special areas and resources: parks, refuges, wilderness, sensitive ecological communities, threatened species of flora and fauna, native communities (including religious sites and harvesting/hunting or subsistence areas).

   - Authority/willingness to require special mitigation: community assistance, staged or phased development, isolate development workforce, pre- and post-development studies and monitoring (with corrective action as needed), worker training, mass transit of workforce.

Natural Resource Issues

Water

6. Bottom disturbance resulting from sampling, platform siting, and pipeline trenching increases particulate dispersion in the water column. In coastal areas, disturbed sediments may contain heavy metals and other contaminants. Production waters are usually more saline than seawater, have little or no dissolved oxygen, and may contain heavy metals, elemental sulfur and sulfides and organic compounds, including hydrocarbons. Discharged drilling muds and additives are contaminated with formation waters and release hydrocarbons, heavy metals, and other contaminants into the water column. Sanitary waste discharges will be highly variable but are usually less diluted than municipal wastes. Routine production activities result in chronic, low level hydrocarbon contamination of the water in the areas around the platforms. Non-routine occurrences such as spills at transfer or loading points, pipeline failure, tanker spill, or well blowout may result in local to widespread severe contamination of the water column.

Air

7. At the drilling and production sites routine emissions include combustion gases from diesel-powered generators and pumps, oil evaporation at transfer and loading points, flaring or venting of waste gas, and small oil spills. Major non-routine emissions can result from catastrophic events such as well blowouts with fire or release of hydrogen sulfide gas, rupture of gas storage tank or transfer line, or evaporation from large oil spills. Transportation-related emissions include barge or tanker product evaporation and fuel combustion, evaporation of oil spills (or discharge of natural gas) from pipeline rupture or ship collision. At the refinery and/or gas processing plant, emissions result from combustion, evaporation and venting during routine operations, and from catastrophic events such as major spills from storage tank rupture or fire.

Land

8. Disturbance of the seafloor can be caused by bottom sampling, anchor dragging, drill ship and platform siting, production facility installation, and pipeline trenching during development. Burial or contamination of bottom features results from discharge of drilling muds, cuttings, and solid wastes. A major oil spill can result in contamination of the sea and coast lines by heavy oil residues. Land disturbance onshore will result from trash and spilled oil floating ashore, clearance of pipeline landings and support facility sites, and secondary effects of increased population.

Sociocultural Issues

Land Use

9. Offshore oil and gas exploration involves temporary non-intensive use of the coastal and offshore areas. Sites required for the offshore production facilities, pipelines, and onshore staging and processing facilities are not available for other uses during the life of the field. Development and production in remote areas will require construction of port facilities and townsites.

Cultural Resources

10. Development and construction may damage or destroy cultural resources, historic sites, or sites of religious significance to native groups. Offshore sites of archeologic importance are particularly vulnerable as they are not readily apparent.

People

11. Drilling and production facilities, vessel traffic, and coastal pipeline landings may interfere with fishing and pleasure boat use of the coastal area. Noise from overflights, near-shore drilling and production operations, port traffic, and processing plant operations will be distracting. The immigration of workers may overtax community services, cause economic, social or cultural conflicts, or even displace local populations, often with "boom and bust" effects. Offshore and onshore facilities have visual impacts. The initial construction force tends to be transient and is soon replaced by the operation staff which is usually smaller and permanent. Control and clean-up from a major oil spill, well blowout or fire, with the resulting rapid marshalling and deployment of large crews, equipment and supplies on an emergency basis, creates a severe but temporary disruption of other activities in the coastal area. Residual effects from a spill would be oil-stained beaches, boats and shore facilities.

Special Issues

Oil Spill Contingency Plans

12. Catastrophic oil spills from well blow-out, pipeline rupture, or tanker/barge collision result in the rapid release of large quantities of oil into the offshore waters, threatening marine mammals, sea and shore birds, and the coastal areas. Spill contingency plans include stockpiling of response equipment, training exercises, and modelling (with local tide and climate data) various spill scenarios. Besides the impacts and disruption of coastal activities occasioned by a major spill, there is the question of compensation for damages (lost fishing revenue, soiled boats and shore structures, loss of recreation benefits and tourist revenues, and natural resource damage and loss). For further discussion, see the section on "Petroleum Refineries."

Oil Refineries and Gas Processing Plants

13. These are major industrial facilities with comparable environmental and human impacts. While the tendency is to site these on or near the coast, they can be located several miles or more inland. Siting considerations include proximity to the offshore facilities, access to existing transportation and community infrastructure and services, space and pipeline routings. Local communities may object to odors, noise, and the risk of fire or gas releases.

Other Issues

14. Other issues that may be of concern include onshore solid waste disposal sites, compatibility with coastal zone management plans, governmental jurisdiction, limits of deep-water technology (for platforms)

and damage to trawls from seafloor litter (dropped or dumped equipment, pipe or rig stubs). Platform siting and design require extensive preconstruction investigation. Important issues of concern include bottom stability, seismic activity, and the probable severity of storms during life of the project. The platforms sometimes attract fish and enhance local sport fishing. At the conculsion of the operation, there may be requests to leave specific ones in place.

Project Alternatives

15. Other than the alternative of "no action" or not going forward with all or parts of the project, alternatives for exploration and production activities/facilities are generally in the type and degree of mitigation that will be required. Special mitigation measures, as amplified in Tables 10.4 and 10.5, may be tailored to the particular project.

16. Pipelines are preferred over barge or tanker for oil and gas transportation (see the discussion on "Oil and Gas Pipelines"). Alternatives for onshore facilities include siting; however, port facilities must be located along the coast. Refineries, gas processing plants, and new community development may be put inland.

Management and Training

17. Basic to safe offshore oil and gas development for protection of workers, the general public, and the environment is an adequate framework of regulations with a competent inspection staff and effective enforcement. Key features to consider at exploration, development, and production sites include installation, inspection and maintenance of all required safety equipment, and training and emergency drills to ensure that workers are experienced in its use. Other key items include blow-out prevention devices, fire suppression equipment, gas detection alarms and hydrogen sulfide protection, platform evacuation devices/procedures, and oil spill response capability.

18. Pressure monitoring and automatic shut-off devices must be installed on pipelines and periodically inspected and tested. Training and safety equipment requirements at the processing facilities are typical of a complex industrial chemical plant, with particular emphasis on fire prevention and suppression, gas detection and control, and spill response.

Monitoring

19. Monitoring requirements include air emissions and waste discharges at production platforms and processing facilities, visual patrols for oil sheen in vicinity of exploration and production sites and along pipeline corridors, and visual patrols for debris and trash (floating in vicinity of operations or washed up on shore). Special monitoring requirements may be established to detect impacts on specific resources in conjunction with mitigation measures. Monitoring at processing facilities includes continual air quality sampling onsite and at site boundaries (automatic samplers), daily visual checks for spillage around tanks, pipelines and transfer points, frequent downstream checks on any surface drainages in the area (visual and water samples), and periodic groundwater sampling onsite and down gradient (monitoring wells).

Table 10.5. Oil and Gas Development—Offshore

Potential Negative Impacts	Mitigating Measures
Direct	

1. Disturbance of cultural resources, benthic communities, coral reefs, coastal barriers, wetlands, pipelines and cables (e.g., anchor dragging, bottom sampling, pipeline trenching, drill ship positioning, platform siting, and so forth).	1. • Require appropriate resource surveys of the offshore and coastal areas that may be affected by the project prior to any disturbance. Typically this will include:
	   • an inventory of cultural and historic resources
	   • an inventory of the flora and fauna of the region
	   • identification of significant topographic features
	   • an inventory of existing offshore pipelines and cables
	
	• Mitigation measures based on identified resource conflicts may include:
	   • avoidance
	   • timing of operations
	   • recovering and archiving cultural and historic resources

2. Degradation of coastal and offshore waters by discharges during routine operations (e.g., drilling muds, sanitary waters, production waters, and spills).	2. • Require separation of cuttings from drilling mud and washing before discharge. • Disposal of drilling muds onshore. • Treatment of formation waters, sanitary and domestic wastes, and cleaning waters/solvents to meet water quality standards before discharge. • Gutters and drip pans, especially at transfer points, to control platform spills.

Table 10.5. Oil and Gas Development--Offshore (continued)

Potential Negative Impacts	Mitigating Measures
Direct (continued)	
	• Water quality standards should be established for all waste water discharges.
	• Drill cuttings checked for residual oil before discharge.
	• Waters in vicinity of platform or drill ship monitored for oil sheen.
3. Degradation of air quality from routine operational emissions (e.g., combustion, venting, spills).	3. • Require appropriate pollution control devices installed and operative on all diesel generators and pumps.
	• Require hydrocarbon vapor control at all oil or gas transfer points, and prompt cleanup of any oil spills.
	• Minimize venting during production.
4. Mortality and/or reduced reproduction of benthic organisms, coral communities, and other marine life through smothering (e.g., disturbed bottom sediments, drill muds, cuttings).	4. • Prohibit or restrict bottom-disturbing activities in vicinity of significant coral reefs and benthic communities.
	• Discharged drill cuttings should be shunted to avoid these features.
	• Spent drilling muds should be barge to shore or discharged well away from any significant live bottom communities.
5. Mortality and/or reduced reproduction of marine flora and fauna, waterfowl, sea birds and waterfowl through oil coating resulting from oil spills.	5. • Minimize routine oil spillage through adherence to water quality, discharge standards, and good housekeeping practices on drill ships, platforms, shuttle boats, barges and tankers, and at transfer points.

Table 10.5. Oil and Gas Development--Offshore (continued)

Potential Negative Impacts	Mitigating Measures
Direct (continued)	
6. Disturbance of marine mammals by seismic surveys, drilling and ship noises.	• Prompt detection and effective response to any operational or catastrophic spills. • Provision for treatment facilities for any oiled birds or aquatic mammals. • Prohibit use of explosives during presence of sensitive marine mammals.
7. Degradation of beach areas, coastal facilities, and boats by oil spills and littering (e.g., coating, tar balls, trash and debris from offshore facilities and transport).	• Solid waste disposal requirements, including sanitary and domestic wastes. • Require labeling of all loose materials and equipment on vessels and platforms (especially barrels, boxes, etc.).
8. Obstruction of boat traffic by offshore facilities.	• Do not site platforms in established shipping lanes.
9. Loss or reduction of fishing areas and recreation sites.	• Do not site platforms in significant fishing or water-oriented recreation areas.
10. Degradation of sea-ward vista (by siting of drilling ships and platforms).	• Paint structures to blend with background (water and sky). • Camouflage structures (however, reducing the visibility of drill ships or platforms may increase navigation hazards). • Use subsurface or bottom production units where feasible.

Table 10.5. Oil and Gas Development—Offshore (continued)

Potential Negative Impacts	Mitigating Measures
Direct (continued)	
11. Congestion and increased boating accidents in the coastal (from increased ship traffic).	11. • Establish and publicize sea-lanes for shuttle traffic. • When possible, avoid areas of heavy recreational or fishing boat use.
12. Disturbance to humans and wildlife by increased noise levels in coastal area from aircraft overflights, ship traffic, and facility operations.	12. Minimize overflights of populated areas.
13. Loss of beach areas to pipeline landfalls and support facilities (e.g., land use, impact of spill cleanup activities, use of dispersants, traffic, disturbance from cleaning activities, and soil contamination).	13. Avoid heavily-used recreational beaches.
14. Injury/loss of life from accidents in transportation and facility operations.	14. • Periodic training and continual safety reminders to all operating staff. • Require periodic drills in emergency procedures. • Ensure that all visitors are briefed on potential hazards and necessary safety precautions. • Ensure that appropriate safety and rescue equipment is available and employees trained in its use. • Install safety valves and alarms in subsurface well-completion systems, with monitoring at production platforms and onshore location.

Table 10.5. Oil and Gas Development—Offshore (continued)

Potential Negative Impacts	Mitigating Measures
Direct (continued)	
15. Contamination of groundwater aquifers (e.g., wells).	15. • Require proper drilling practices, casing, and sealing off all aquifers during drilling. • Ensure that all aquifers are properly sealed off prior to well completion or abandonment.

16. Increased demands on community facilities and services in the coastal area.	16. • Require pre-development, socio-economic studies of potentially affected communities to identify possible impacts on services, infrastructure, dislocations, and conflicts.
	
	• These impacts can be addressed by:
	  • community assistance grants
	  • loans
	  • pre-payment of taxes
	  • phasing the oil and gas development
	  • constructing needed community facilities
	
	• Cooperative and open working relations should be established early with local communities and maintained throughout the life of the project.
	
	• Project workers should be encouraged to participate in community affairs.

Table 10.5. Oil and Gas Development—Offshore (continued)

Potential Negative Impacts	Mitigating Measures
Direct (continued)	
17. Conflicts with native cultures, traditions, and life-styles.	17. • Brief all employees to ensure awareness of and sensitivity to the local cultures, traditions, and lifestyles. • Ensure that native leaders are aware of the projected activities, are assisted in identifying impacts that may be of particular concern to them, and have a voice in appropriate mitigation measures. • Mitigation may include isolating the development work force from the native community.

OIL AND GAS DEVELOPMENT--ONSHORE

1. This category includes prospecting, exploration, development and production of onshore oil and gas resources. Typically, geological and geophysical studies are conducted over large areas to identify favorable exploration targets. This is followed by more intensive study, testing and drilling in selected areas to locate and evaluate the oil and gas resource. Production facilities include wells and pumps spaced over the field, gathering and transportation lines, storage tanks, and some primary processing units. World Bank projects range from the initial exploration of large regions to development of production facilities for oil, natural gases, and byproducts. Production projects can include secondary or enhanced recovery.

Potential Environmental Impacts

2. Roads into undeveloped areas result in surface disturbance, traffic noises, and increased access. Seismic exploration involves noise and disturbance from explosive charges either in shallow shot holes or above ground. Geologic test or exploratory wells require intensive surface disturbance for the well site, access roads, air field, noise from truck or air transport traffic, construction and operation, air emissions from transport traffic and drilling operations, and discharge of drilling fluids contaminated with mud additives, formation waters, and oil. Access roads, seismic operations, and exploratory wells have the potential to damage cultural resources and sensitive ecosystems or adversely affect native communities, if located or designed improperly. When these activities are properly planned, designed and conducted, as amplified in Table 10.6, the impacts should be temporary.

3. Oil and gas production requires extensive on-site industrial activities for the life of the field. Construction of wellsites, access roads, air field(s), gathering and transport pipelines, and ancillary support facilities will result in extensive surface disturbance, construction traffic, noise and air emissions, and influx of construction crews. Production from small fields and initial production from large fields may be trucked to refineries, adding to vehicular traffic, accidents, and oil spillage. In remote areas, facilities will be required for the permanent operation and maintenance staff. Production operations limit other land uses of the area. There will be continual noise and air emissions from equipment operations, discharge of treated formation waters, and oil spills. Air pollution may also result from flaring of unwanted gases, sour gas discharge (hydrogen sulfide), and burning of oil waste pits. (See other examples in Table 10.7 at the end of this section.)

4. Potential catastrophic accidents include well blow-out with uncontrolled oil and/or gas release and perhaps fire (with combustion products), refinery or gas processing plant fire.

Natural Resource Issues

Water

5. Contamination of local surface waters can result from improper handling of drilling fluids and production water, leakage from pipelines, pits and storage tanks, rainwater runoff from roads, pads and

Table 10.6. Checklist--Onshore Oil and Gas Development

1. **Production**

   - Production Zone(s): Reserves, depth, area, structure, oil/gas/water ratios, oil type, gas type(s), pressures.

   - Operations: Site preparation, well spacing, start-up period (production rate, product transportation, field life, waste sanitary wastes), pollution control, monitoring, spill response and reclamation plans.

   - Air Emissions: Emission quantity and, where applicable, composition: dust, venting, flaring, waste pit burning, combustion, equipment emissions, oil spills and leakage.

   - Waste Water Discharge: Projected quantity and composition; treatment/disposal method (production water, sanitary wastes).

   - Services: Utilities (type, source, load), roads, airfields, rail, fire protection, security.

   - Landuse: Area of field, transportation and utility routes, pipelines, buildings and structures (at field and stations along pipeline).

   - Equipment: Type and number for site preparation, drilling, production, transportation, waste water separations and disposal, waste haulage, pumping, reclamation, transport of supplies and workers.

   - Supplies: Drilling muds, pipe, chemicals, water, fuel.

   - Staffing: Construction, production, reclamation phases, number and skills, source, housing plans.

2. **Environmental Resources**

   - Geology: Stratigraphy, structure, fracture patterns, seismic history.

   - Groundwater: Depth to and thickness of aquifers, quality and quantity, hydraulics, recharge, uses.

   - Surface Water: Quality, quantity, seasonal variations, uses.

   - Soil: Soil profile (depth, type, characteristics).

   - Vegetation: Types, density, rare or significant species or communities, wetlands.

   - Wildlife: Resident and casual, populations, rare or significant species, significant habitat.

   - Topography: Drainage patterns, elevations and slopes, prominent features.

   - Climate: Precipitation patterns (amount, frequency, type), air quality, wind patterns (direction, speed, frequency), temperature, climatic zone.

3. **Socio-Economic Factors**

   - Nearby communities: Location, access, population (number, demographic and social characteristics); economy (employment rate, income distribution, tax base); services (types, capacity, adequacy) and housing; concern is the ability to (a) provide workforce, (b) service new development and (c) absorb and adjust to growth (worker/family in-migration).

   - Land Use: Intensive and casual, full time and seasonal, actual and projected, specially designated areas (parks, refuges, reservations, wilderness), man-made features (structures, roads, utilities).

   - Cultural: Historic sites, archaeologic sites, native religious or harvest sites.

4. **Regulatory Framework**

   - Applicable environmental laws, regulations, policies, standards, and requirements; monitoring and enforcement: air, water, waste, noise, reclamation, land use controls and approvals, cultural and historic resource protection.

   - Designation and protection of special areas and resources: parks, refuges, wilderness, sensitive ecological communities, threatened species of flora and fauna, native communities (including religious sites and harvesting/hunting or subsistence areas).

   - Authority/willingness to require special mitigation: community assistance, staged or phased development, isolate development workforce, pre- and post-development studies and monitoring (with corrective action as needed), worker training, mass transit of workforce.

other paved or packed surfaces, improper handling of domestic sewage and wastes from equipment maintenance and erosion of disturbed soils. Water needed for drilling and domestic purposes may be taken from local sources, thus diminishing that available for local natives or wildlife. Improper casing of wells can lead to contamination of aquifers. Accidents, such as rupture of a pipeline or storage tank, may result from improper installation or poor maintenance, aging facilities, third-party actions (sabotage, collisions, etc.) and seismic occurrences (soil slumpage). Infrequently, unexpected drilling conditions can result in a well blow-out, causing an uncontrolled release of large volumes of oil and/or gas and formation waters into the surface waters.

Air

6. Air-borne particulates will result from soil disturbance during construction activities and from vehicle traffic and wind erosion on unpaved roads and other disturbed surfaces. Other air pollutants, as well as particulates, will result from burning of sludge pits and various wastes and any flaring of gas. Hydrocarbon emissions will result from system venting, any leaks or spills, and residue in production wastes. Vehicles and any gasoline or diesel-powered facilities will emit CO, NO_x, etc. Where small amounts of gas are produced along with oil, these may be vented or flared (burned from a stack). Emissions from this source could include sour gas (H_2S), CO_2, methane, etc. A well blow-out could result in a large-scale uncontrolled release of natural gas or H_2S or fire (with large-scale emission of NO_x, SO_x, CO, and TSP).

Land

7. Construction of roads, airfields, drill sites, and production facilities involves the removal of vegetation and usually some modification of topography. Clearing of seismic lines may be required in areas of dense vegetation. Combustible vegetation will be controlled for the life of the project around wells, pipelines and storage tanks, and ancillary production facilities. Some modification may result through planting of non-native species for erosion control. Reduction or modification of vegetation can reduce livestock forage, wildlife habitat, and timber yield. Sensitive ecological areas, critical wildlife habitat, and endangered plant species can be inadvertently damaged or destroyed. Wetlands may be modified by access roads and drill pads, drained or filled. Habitat reduction or loss and on-going human activities will lead to some loss in wildlife populations, and sensitive species may be eliminated from the area.

Sociocultural Issues

8. The prospecting and exploration are short-term non-intensive land uses which will have minimal effect on existing non-oil and gas activities. In remote areas, these activities can be serviced by air, thus eliminating the need (and intrusion) of access roads. Production activities, on the other hand, will constrain other uses of the developed area for the life of the field. Production in remote areas will involve new access roads and townsite(s).

9. Construction and other land disturbance activities may inadvertently damage or destroy cultural resources, historical sites, or sites of religious significance to native groups. Increased human presence in the area may lead to vandalism of unprotected sites.

10. Drilling and production, transportation, and processing facilities will interfere to some degree with other activities in the area. Noise from truck traffic and aircraft overflights, drilling, pumps, and processing plants will be distracting. The immigration of workers may overtax community services and cause economic, social or cultural conflicts, or displace local populations. There is the potential for a "boom and bust" phenomenon. The initial exploration and construction force is usually transient and soon replaced by a smaller, permanent operation staff. Control and cleanup from a major oil spill, well blow-out, or fire requires rapid marshalling and deployment of large crews, equipment, and supplies on an emergency basis and creates a severe but temporary disruption of nearby communities and other activities in the area.

Special Issues

Coastal Areas and Wetlands

11. Exploration in wetlands, coastal marshes, and shallow coastal waters may be carried out with minimum disruption by use of helicopter staging and/or temporary roads. Production will require an access road into the field, well pads, separators and storage tanks, utilities, gathering lines and a product pipeline. Special concerns are modification of the wetland hydrologic regime by diking from roads and pipeline corridors, degradation of water quality and vegetation in these highly productive areas from leaks or spills of drilling muds, oil, water production, and loss of wetlands area to roads and drill/production pads. In shallow coastal waters, causeways to production sites may block fish movement lateral to the coastline unless openings are provided, and current modifications may affect local beaches.

Sensitive Areas and Wilderness

12. Exploration in remote areas by aerial access to avoid the need for access roads requires many trips by large helicopter (or cargo plane if landing field available). The noise of overflights can be disturbing to sensitive wildlife and detract from wilderness experience of visitors. Production will require road, pipeline, and utility access to the area; however, improved access may encourage encroachment and induce development. This can result in conflicts with native culture, disruption of sensitive wildlife habitat, or loss of wilderness values.

Project Alternatives

13. Other than the alternative of "no action" or not going forward with all or parts of the project, alternatives for exploration and production activities/facilities are generally in the type and degree of mitigation that will be required. Special mitigation measures may be tailored to the particular project. Because pipelines are generally preferred over trucks for product transportation, alternatives should include routing options. For processing facilities, alternatives should include siting options.

Management and Training

14. Basic to safe onshore oil and gas development, for the protection of workers, the general public and the environment, are adequate regulations with a competent inspection staff and effective enforcement.

Key features at exploration, development, and production sites include installation, inspection and maintenance of all required safety equipment, training workers in its use through instruction and emergency drills. Key items that should be considered include blow-out preventers, fire suppression equipment, gas detection alarms, hydrogen sulfide protection, rig evacuation lines, and spill response capability. Pressure monitoring and automatic shutoff devices must be installed on pipelines and periodically inspected and tested. Training and safety at the processing facilities is typical of a complex industrial chemical plant with particular emphasis on fire suppression, gas detection and control, and spill response.

Monitoring

15. Monitoring requirements include air emissions and waste discharges at drill rigs and processing facilities, visual checks of integrity of sludge pits and tank dikes, visual checks of areas around wells, pipeline routes, storage tanks, and transfer points for spills or leakage. Special monitoring requirements may be imposed for early detection of impacts on specific resources in conjunction with mitigation measures. Monitoring at processing facilities includes continual air quality sampling onsite and at site boundaries (automatic samplers), daily visual checks for spillage around tanks, pipelines and transfer points, frequent downstream checks of any surface drainages in the area (visual and water samples), and periodic groundwater sampling onsite and down gradient (monitoring wells).

Table 10.7. Oil and Gas Development--Onshore

Potential Negative Impacts	Mitigating Measures
Direct	

1. Disturbance of cultural resources, historic sites, vegetation, wetlands, surface drainages, and wildlife.	1. • Require appropriate resource surveys of the areas that may be affected by the project prior to any disturbance. Typically this will include:
	  • an inventory of cultural and historic resources
	  • an inventory of the flora and fauna of the region
	  • identification of significant topographic features
	  • an inventory of existing pipelines and cables
	
	• Mitigation measures based on identified resource conflicts may include:
	  • avoidance
	  • timing of operations
	  • recovering and archiving cultural and historic resources
	  • compensating for losses by protecting or enhancing comparable resources in region

2. Degradation of surface waters by soil erosion from disturbed areas, discharge of drilling slurries and produced waters, wastes equipment servicing, and sanitary and domestic wastes.	2. • Require control of stormwater runoff and prompt revegetation on disturbed areas. • Burial of drilling muds and cuttings. • Re-injection of produced formation waters. • Good housekeeping practices at drill and production sites to minimize leaks and spills.

Table 10.7. Oil and Gas Development—Onshore (continued)

Potential Negative Impacts	Mitigating Measures
Direct (continued)	• Treatment of sanitary/domestic wastes and cleaning waters/solvents to meet water quality standards before discharge.
	• Prompt cleanup of any spills (oil, drilling mud, formation water).
	• Water quality standards should be established for all waste water discharges.
3. Degradation of air quality from routine operational emissions.	3. • Require appropriate pollution control devices on diesel generators and pumps, and hydrocarbon vapor control at all oil or gas transfer points.
	• Require prompt cleanup of any oil spills.
	• Minimize venting during production.
4. Mortality and reduced reproduction of wildlife from habitat disturbance or loss, road kills, and hunting.	4. • Prohibit or restrict disturbance of significant habitat and wetlands.
	• Mark wildlife road crossings.
	• Prohibit firearms possession in the area.
5. Modification of vegetation and introduction of non-native species.	5. Require prompt reclamation of disturbed areas and revegetation with native species.

Table 10.7. Oil and Gas Development—Onshore (continued)

Potential Negative Impacts	Mitigating Measures
Direct (continued)	
6. Degradation/loss of vegetation (and soil productivity) from discharge or spills of produced waters, oil, and drilling muds.	6. • Require blow-out preventers and control and prompt cleanup of oil and formation water spillage. • Keep soil disturbance and vegetation clearing to minimum required for operation and safety.
7. Land-use conflicts.	7. • Consult with local land users in siting access, air fields, utility lines, and, to extent possible, production facilities. • Allow other land uses to continue on the site where compatible with the operations.
8. Degradation of remote areas through improved access and increased use.	8. • Access remote areas by air during early exploration stage. • Restrict use of access roads. • Remove and reclaim any access roads at end of production. • Minimize need for community development by rotating work crews and precluding permanent residence.
9. Road damage, accidents, and traffic delays from increased truck traffic on local roads.	9. • Observe road load limits. • Design roads for adequate capacity and visibility. • Ensure that roads are properly signed, vehicles are well-maintained, and drivers are trained and safety-conscious. • Require that commuting workers car-pool or provide buses.

Table 10.7. Oil and Gas Development--Onshore (continued)

Potential Negative Impacts	Mitigating Measures
Direct (continued)	
10. • Visual intrusions from wells, tanks, and production facilities.	10. • Paint structures to blend with background (vegetation and sky).
• Cleared linear rights-of-way for pipelines, utilities, and roads, and processing facilities.	• Avoid contrasting colors.
	• Utilize utility corridors.
	• Minimize clearing and blend vegetation where feasible.
11. Disturbance of humans and wildlife by noise from seismic surveys, drilling, pumping, and processing facilities.	11. • Avoid seismic shots, low overflights, and other sudden loud noises in critical wildlife areas, especially during mating or nesting season.
	• Require proper mufflers on diesel equipment.
12. Loss of birds and animals in sludge ponds.	12. • Minimize surface area and number of sludge pits, and require that they be promptly drained, closed, or covered (with netting) when not in use.
13. Injury/loss of life from accidents.	13. • Periodic training and continual safety reminders to all operating staff.
	• Require periodic drills in emergency procedures.
	• Ensure that all visitors are briefed on potential hazards and necessary safety precautions.
	• Ensure that appropriate safety and rescue equipment is available and employees trained in its use.
	• Install subsurface safety valves on gas producing wells.

Table 10.7. Oil and Gas Development—Onshore (continued)

Potential Negative Impacts	Mitigating Measures
Direct (continued)	
14. Contamination of groundwater aquifers.	14. • Require proper drilling practices, casing and sealing off all aquifers during drilling. • Ensure that all aquifers are properly sealed off prior to well completion or abandonment. • Line all mud storage and waste fluid pits.

15. Increased demands on services and facilities in local communities, social and cultural conflicts, concern with community stability (boom and bust scenario).	15. • Require pre-development, socio-economic study of potentially affected communities to identify possible impacts on services, infrastructure, dislocations, and conflicts.
	
	• These impacts can be addressed by:
	   • community assistance grants
	   • loans
	   • pre-payment of taxes
	   • phasing the oil and gas development
	   • constructing needed community facilities
	
	• Cooperative and open working relations should be established early with local communities and maintained throughout the life of the project.
	
	• Project workers should be encouraged to participate in community affairs.

Table 10.7. Oil and Gas Development--Onshore (continued)

Potential Negative Impacts	Mitigating Measures
Direct (continued)	
16. Conflicts with native cultures, traditions, and life-styles.	16. • Brief all employees to ensure awareness of and sensitivity to the local cultures, traditions, and lifestyles. • Ensure that native leaders are aware of the projected activities, are assisted in identifying impacts that may be of particular concern to them, and have a voice in appropriate mitigation measures. • Mitigation may include isolating the development work force from the native community.
17. Subsidence of land surface.	17. • Re-injection of produced formation water and injection of additional water to replace volume of oil removed.
18. Use of local surface water or groundwater.	18. • Obtain water from unutilized aquifers. • Non-potable water can be used for drilling, sprinkling roads, and irrigating.

HYDROELECTRIC PROJECTS

1. Hydroelectric projects include dams, reservoirs, canals, penstocks, powerhouses, and switchyards for the generation of electricity. The dam and reservoir may be multi-purpose; if watershed rainfall and stream flow characteristics and water and power usage patterns permit, hydroelectric reservoirs can also provide one or more of the following services: irrigation, flood control, water supply, recreation, fisheries, navigation, sediment control, ice jam control, and glacial lake outburst control. However, these are competing uses for the waters stored behind dams and each may imply a different diurnal or annual operating rule curve for the reservoir.

2. In a hydroelectric project, for example, the operator will maximize power benefits by varying reservoir level in accordance with a rule curve that is close to the reservoir trajectory in a very dry year. For flood control, an operator will draw down the reservoir to have maximum volume for flood retention available at the beginning of the rainy season. Irrigation reservoirs will be filled and releases made according to the growing seasons of the crops being irrigated. In project planning and development of the rule curve, any conflicts between competing uses should be resolved.

3. Hydroelectric projects necessarily entail construction of transmission lines to convey the power to its users. Electric power transmission lines are covered separately in the section on "Electric Power Transmission Systems."

Potential Environmental Impacts

4. The principal source of impacts in a hydroelectric project is the construction and operation of a dam and reservoir (see the section on "Dams and Reservoirs"). Large dam projects cause irreversible environmental changes over a wide geographic area and thus have the potential for significant impacts. Criticism of such projects has grown in the last decade. Severe critics claim that the social, environmental, and economic costs of dams outweigh their benefits and that the construction of large dams, therefore, is unjustifiable. Others contend that in some cases environmental and social costs can be avoided or reduced to an acceptable level by carefully assessing potential problems and implementing cost-effective corrective measures.

5. The area of influence of a dam and its reservoir extends from the upper limits of the reservoir to as far downstream as the estuarine, coastal and offshore zones, and includes the reservoir, dam and river valley below the dam. While there are direct environmental impacts associated with the construction of the dam (e.g., dust, erosion, borrow and disposal problems), the greatest impacts result from the impoundment of water, flooding of land to form the reservoir, and alteration of water flow downstream. These effects have direct impacts on soils, vegetation, wildlife and wildlands, fisheries, climate and human populations in the area. (See Table 10.8 at the end of this section for other examples and recommended mitigation measures.)

6. The dam's indirect effects include those associated with the building, maintenance and functioning of the dam (e.g., access roads, construction camps, power transmission lines) and the development of agricultural, industrial or municipal activities made possible by the dam.

7. In addition to the direct and indirect effects of dam construction on the environment, the effects of the environment on the dam must be considered. The major environmental factors affecting the functioning and lifespan of the dam are those caused by land, water, and other resource use in the catchment areas above the reservoir (e.g., agriculture, settlement, forest clearing) which may result in increased siltation and changes in water quantity and quality in the reservoir and river downstream. These are usually thoroughly addressed in the engineering studies.

8. The obvious benefit of a hydroelectric project is electric power, which can support economic development and improve the quality of life in the service area. Hydroelectric projects are labor-intensive, as well as providing employment opportunities. Roads and other infrastructure may provide local inhabitants with better access to markets for their crops, educational facilities for their children, health care, and other social services. Moreover, the generation of hydropower provides an alternative to the burning of fossil fuels or to nuclear power, which allows the power demand to be met without producing heated water, air emissions, ash, and radioactive waste. If the reservoir is a truly multi-purpose facility, that is, if the various purposes stated for it in the economic analysis are not mutually inconsistent, other benefits may include flood control and the provision of a more reliable and higher quality water supply for irrigation, domestic and industrial use. Intensification of agriculture locally through irrigation can in turn reduce pressure on uncleared forest lands, intact wildlife habitat, and areas unsuitable for agriculture elsewhere. In addition, dams create a reservoir fishery and the possibilities for agricultural production on the reservoir drawdown area, which in some cases can more than compensate for losses in these sectors due to dam construction.

Special Issues

Hydrologic and Limnological Effects

9. Damming the river and creating a lake-like environment has profound effects on the hydrology and limnology of the river system. Dramatic changes occur in the timing of flow: quality, quantity and use of water, aquatic biota, and sedimentation dynamics in the river basin. Hydroelectric projects, in particular, are liable to create major changes in the river flow patterns downstream because storage and releases are managed in response to power demand cycles rather than the hydrologic cycles to which the downstream riparian environment is adapted.

10. The decomposition of organic matter on the flooded lands creates a nutrient-rich environment. Fertilizers used upstream may add to the nutrients which accumulate and recycle in the reservoir. This not only supports an active reservoir fishery but stimulates the growth of aquatic weeds, such as water lettuce and water hyacinth. Weeds and algal mats can be expensive nuisances which clog dam outflows and irrigation canals, negatively affect fisheries and recreation, increase water treatment costs, impair navigation, and substantially increase water loss through transpiration.

11. If the inundated land is heavily wooded and not sufficiently cleared prior to flooding, decomposition will deplete oxygen levels in the water. This affects aquatic life and may result in large fish kills. Products of anaerobic decomposition include hydrogen sulfide, which corrodes generator turbines and is noxious to aquatic organisms and methane, which is explosive and a greenhouse gas.

12. Oxygen depletion typically occurs first in the deeper water where oxygen used by bacteria in decomposition is not offset by oxygen released by plant photosynthesis. When the intake for power generation is located at the lower level of the reservoirs, as is usually the case, water released from the turbines into downstream waters may be deficient in oxygen and may contain hydrogen sulfide. It may also be lower in pH and colder than the surface water. Releases of water with these characteristics can adversely affect plant and animal communities in the river below the dam.

13. Suspended particles carried by the river settle in the reservoir, limiting its storage capacity and lifetime, and robbing downstream waters of sediment. Many agricultural areas on floodplains have always depended on nutrient-rich silt to sustain productivity. As sediment is no longer deposited on the floodplain downstream, the loss of nutrients must be compensated by fertilizer to maintain agricultural productivity. The release of relatively sediment-free waters can result in the scouring of the downstream riverbeds (which may be beneficial in some cases, detrimental in others).

14. Additional effects of changes in the hydrology of the river basin include altered levels of the water table both above and below the reservoir, and upstream encroachment of saline waters into estuaries that have direct ecological impacts and affect downstream water users.

Social Issues

15. The benefits of hydroelectric dams often accrue to communities and industries some distance away from the dam. Those who typically bear the heaviest environmental and social costs of the dam construction -- namely, the inhabitants of area inundated by the reservoir -- may not receive a proportionate share of the benefits. Reservoir filling results in the forced relocation of those living on the land (in some projects from hundreds of thousands to over a million people), which profoundly affects them as well as those present in the resettlement areas. (For information on "Involuntary Resettlement," see Chapter 3). For those remaining in the river basin, access to water, land and biotic resources often is restricted. Artisanal riverine fisheries and traditional floodplain (recession) agriculture are disrupted by changes in stream flow and reduced silt deposition. Floodplains of many tropical rivers are vast areas of great importance for human and animal populations. When the floodplains shrink, land use must change or populations are forced to move. Water-related diseases (e.g., malaria, schistosomiasis, onchocerciasis, encephalitis) may increase with the creation of a reservoir and associated water management structures, if they are endemic to the area.

16. Social and environmental problems arise from controlled and uncontrolled influx of population groups into the area, such as construction workers and power plant employees, seasonal laborers for other dam-induced activities, and rural people who take advantage of increased access to the area provided by roads, transmission lines or improved river navigation (see "New Land Settlement" and "Induced Development" sections in Chapter 3). The consequences are health problems, overburdened public services, competition for resources, social conflicts, and negative environmental impacts on the watershed, reservoir and downstream river basin.

Fisheries and Wildlife

17. As mentioned, riverine fisheries can be expected to decline because of changes in river flow, deterioration of water quality, loss of spawning grounds and barriers to fish migration. However, a reservoir fishery, sometimes more productive than the previous riverine fishery, is created.

18. In rivers with biologically productive estuaries, both marine and estuarine fish and shellfish suffer from changes in water flow and quality. Changes in freshwater flows and thus the salinity balance in an estuary will alter species distribution and breeding patterns of fish. Changes in nutrient levels and a decrease in the quality of the river water can also have profound impacts on the productivity of an estuary. These changes can also have major effects on marine species which feed or spend part of their life cycle in the estuary or are influenced by water quality changes in the coastal areas.

19. The greatest impact on wildlife will come from loss of habitat resulting from reservoir filling and land-use changes in the watershed. Migratory patterns of wildlife may be disrupted by the reservoir and associated developments. Poaching and eradication of species considered to be agricultural pests have a more selective effect. Aquatic fauna, including waterfowl, reptiles, and amphibian populations are expected to increase on the reservoir.

Seismic Threat

20. Considerable evidence has linked the creation of reservoirs with seismic events, but the probability that the reservoir will induce seismicity in aseismic areas is difficult to predict. In seismic areas, a reservoir may advance the occurrence of an earthquake, possibly resulting in more frequent but less destructive events. The EA should consider both of these phenomena.

Watershed Management

21. Increased pressure on upland areas above the dam is a common phenomenon caused by the planned resettlement of people from the inundated areas and by the uncontrolled influx of new people into the watershed. On-site environmental deterioration, as well as a decrease in water quality and increase in sedimentation rates in the reservoir, results from clearing of forest land for agriculture, grazing pressures, use of agricultural chemicals, and tree cutting for timber or fuelwood. Similarly, land use in the watershed of the lower river basin affects the quality and quantity of water entering the river. It is essential that dam projects be planned and managed in the context of overall river basin and regional development plans, including both the upland catchment areas above the dam and floodplain, and the watershed areas downstream.

Project Alternatives

22. A variety of alternatives to proposed hydroelectric projects may exist. Individually or in combination, they may influence size, location, and timing of a proposed hydroelectric project.

- Alter demand for energy by conservation measures, efficiency improvements, or restrictions on regional growth.

- Utilize thermal power plants or alternative energy sources, including low-head hydropower cogeneration by industry, biogas, etc.

- Investigate possibilities for siting the project on an already dammed river by diversifying the functions of that dam.

- Site the proposed project on the river where it will minimize the negative and social impacts.

- Adjust dam height, inundation area, and dam design to minimize negative environmental impacts.

Management and Training

23. Responsibility for management of a single-purpose hydroelectric project is typically the responsibility of a public or private electric utility. Multi-purpose reservoirs may be managed or overseen by government agencies with broader authority to allocate water to competing uses, although it is not uncommon in some countries for a power utility to provide and manage recreational access areas for the public and to sell excess water to other users. Whatever the arrangement, it is important that there be provision for coordinated planning of land and water use in the watershed and region. Often this is accomplished by a river basin authority or other regional entity. The section on "Dams and Reservoirs" provides more information on this topic.

24. The organization operating the dam and reservoir should be in charge of collecting baseline data, building and managing the dam, producing a water-use master plan with management strategies for regulation of the reservoir, and controlling disease vectors. It has become a common practice, and is advisable in most cases, for the operating entity to establish an environmental management unit at the project level. The operating organization should also be consulted in planning for municipal water supplies and water treatment facilities, as well as notified and asked to comment on permit requests for major water withdrawals and wastewater discharges upstream.

25. The line agency with responsibility for energy should ensure intersectoral cooperation at both the policy and field levels with government ministries responsible for agriculture, fisheries, forestry, range and livestock, health, wildlife, tourism, municipal and industrial planning, and transportation.

26. An environmental and economic monitoring program should be established. The monitoring work could be carried out by the implementing agency and/or the operating organization, or by the river basin authority, if one exists. Disciplines which may need to be involved in designing the program and interpreting results include hydrology, limnology, fisheries, forestry/botany, wildlife ecology, livestock/range management, rural sociology and health.

27. To take maximum advantage of training opportunities, project environmental staff should be in place early and should be involved in the EA, the development of mitigation measures and monitoring program, and supervision of construction. They are then in the best position to understand the environmental aspects of the project and to carry out monitoring and environmental management.

Monitoring

28. There is no standard monitoring program for a hydroelectric project, but the EA should include one that is designed for the specific project. Which ones of the following variables to monitor depends on the management information needs.

- rainfall
- stored water volume in the reservoir
- annual volume of sediment transported into reservoir
- water quality at dam discharge and at various points along the river, including:
  - salinity
    - pH
    - temperature
    - electrical conductivity
    - turbidity
    - dissolved oxygen
    - suspended solids
    - phosphates
    - nitrates
- river flow at various points downstream
- volume of water used, by use type, at the reservoir and downstream
- hydrogen sulfide and methane generation behind dam
- limnological sampling of microflora, microfauna, aquatic weeds and benthic organisms
- fisheries assessment surveys (species, population size) in the river and reservoir
- wildlife (species, distribution, numbers)
- livestock (species, numbers, distribution, condition)
- vegetation changes (cover, species composition, growth rates, biomass) in the upper watershed, reservoir drawdown zone, and downstream areas
- impacts on wildlands, species or plant communities of special ecological significance
- public health and disease vectors
- in- and out-migration of people to area
- changes in economic and social status of resettlement populations and people remaining in the river basin

Table 10.8. Hydroelectric Projects

Potential Negative Impacts	Mitigating Measures
Direct	
1. • Negative environmental effects of construction: • air and water pollution from construction and waste disposal • soil erosion • destruction of vegetation • sanitary and health problems from construction camps	1. • Measures to minimize impacts: • air and water pollution control • careful location of camps, buildings, borrow pits, quarries, spoil and disposal sites • precautions to minimize erosion • land reclamation
2. Dislocation of people living in inundation zone.	2. • Relocation of people to suitable area. • Provision of compensation in kind for resources lost. • Provision of adequate health services, infrastructure, and employment opportunities.
3. Loss of land (agricultural, forest, range, wetlands) by inundation to form reservoir.	3. • Siting of dam to decrease losses. • Decrease of dam and reservoir size. • Protection of equal areas in region to offset losses. • Creation of useable land in previously unsuitable areas to offset losses.
4. Loss of historic, cultural or aesthetic features by inundation.	4. • Siting of dam or decrease of reservoir size to avoid loss. • Salvage or protection of cultural properties.

Table 10.8. Hydroelectric Projects (continued)

Potential Negative Impacts	Mitigating Measures
Direct (continued)	
5. Loss of wildlands and wildlife habitat.	5. • Siting of dam or decrease of reservoir size to avoid/minimize loss. • Establishment of compensatory parks or reserved areas. • Animal rescue and relocation.
6. Proliferation of aquatic weeds in reservoir and downstream impairing dam discharge, irrigation systems, navigation and fisheries and increasing water loss through transpiration.	6. • Clearance of woody vegetation from inundation zone prior to flooding (nutrient removal). • Weed control measures. • Harvest of weeds for compost, fodder or biogas. • Regulation of water discharge and manipulation of water levels to discourage weed growth.
7. Deterioration of water quality in reservoir.	7. • Clearance of woody vegetation from inundation zone prior to flooding. • Control of land uses, wastewater discharges, and agricultural chemical use in watershed. • Limit retention time of water in reservoir. • Provision for multi-level releases to avoid discharge of anoxic water.

Table 10.8. Hydroelectric Projects (continued)

Potential Negative Impacts	Mitigating Measures
Direct (continued)	
8. Sedimentation of reservoir and loss of storage capacity.	8. • Control of land use in watershed (especially prevention of conversion of forests to agriculture). • Reforestation and/or soil conservation activities in watersheds (limited affect). • Hydraulic removal of sediments (flushing, sluicing, release of density currents). • Operation of reservoir to minimize sedimentation (entails loss of power benefits).
9. Formation of sediment deposits at reservoir entrance creating backwater effect and flooding and waterlogging upstream.	9. Sediment flushing, sluicing.
10. Scouring of riverbed below dam.	10. Design of trap efficiency and sediment release (e.g., sediment flushing, sluicing) to increase salt content of released water.
11. Decrease in floodplain (recession) agriculture.	11. Regulation of dam releases to partially replicate natural flooding regime.
12. Salinization of floodplain lands.	12. Regulation of flow to minimize effect.
13. Salt water intrusion in estuary and upstream.	13. Maintenance of at least minimum flow to prevent intrusion.

Table 10.8. Hydroelectric Projects (continued)

Potential Negative Impacts	Mitigating Measures
Direct (continued)	
14. Disruption of riverine fisheries due to changes in flow, blocking of fish migration, and changes in water quality and limnology.	14. • Maintenance of at least minimum flow for fisheries. • Provision of fish ladders and other means of passage. • Protection of spawning grounds. • Aquaculture and development of reservoir fisheries in compensation.
15. Snagging of fishing nets in submerged vegetation in reservoir.	15. Selective clearance of vegetation before flooding.
16. Increase of water-related diseases.	16. • Design and operation of dam to decrease habitat for vector. • Vector control. • Disease prophylaxis and treatment.
17. Conflicting demands for water use.	17. • Planning and management of dam in context of regional development plans. • Equitable allocations of water between large and small holders and between geographic regions of valley.
18. Social disruption and decrease in standard of living of resettled people.	18. • Maintenance of standard of living by ensuring access to resources at least equalling those lost. • Provision of health and social services.

Table 10.8. Hydroelectric Projects (continued)

Potential Negative Impacts	Mitigating Measures
Direct (continued)	
19. Environmental degradation from increased pressure on land.	19. • Choice of resettlement site to avoid surpassing carrying capacity of the land. • Increase of productivity or improve management of land (agricultural, range, forestry improvements) to accommodate higher population.
20. Disruption/destruction of tribal/indigenous groups.	20. Avoid dislocation of unacculturated people and where not possible, relocate in area allowing them to retain lifestyle and customs.
21. Increase in humidity and fog locally, creating favorable habitat for insect disease vectors (mosquitos, tsetse).	21. Vector control.
Indirect	
22. Uncontrolled migration of people into the area made possible by access roads and transmission lines.	22. Limitation of access, provision of rural development, and health services to try to minimize impact.
23. Environmental problems arising from development made possible by dam (irrigated agriculture, industries, municipal growth).	23. Basin-wide integrated planning to avoid overuse, misuse, and conflicting uses of water and land resources.
External	
24. Poor land use practices in catchment areas above reservoir resulting in increased siltation and changes in water quality.	24. Land use planning efforts which include watershed areas above dam.

THERMOELECTRIC PROJECTS

1. Bank-supported thermoelectric power projects may include gas-fired steam, oil-fired steam, coal-fired steam, combined cycles, gas turbine, and diesel power plants. (The Bank has also participated in geothermal energy projects and may finance solar energy and alternative fuel projects, but these are currently uncommon and are not discussed in this section.) The major components of thermoelectric projects include the power system (i.e., power source turbine and generator) and associated facilities, which may include the cooling system, stack gas cleaning equipment, fuel storage and handling areas, fuel delivery systems, solid waste storage areas, worker colonies, electrical substations and transmission lines. The type of facility and size of thermoelectric projects, as well as location, will determine the type and size of these associated facilities.

Potential Environmental Impacts

2. Negative impacts can occur both during construction and operation of thermoelectric plants. Construction impacts are caused primarily by the following site preparation activities: clearing, excavation, earth moving, dewatering, dredging and/or impounding streams and other water bodies, establishing lay-down areas, and developing borrow and fill areas. The large number of workers employed in constructing power plants can have significant sociocultural impacts on local communities.

3. Thermoelectric plants are considered major air emission sources which can affect local and regional air quality. Sulfur dioxide (SO_2), oxides of nitrogen (NO_x), carbon monoxide (CO), carbon dioxide (CO_2) and particulates (which may contain trace metals) are emitted from the combustion of fuels by thermoelectric projects. The amounts of each depend on the type and size of facility, the type and quality of fuel, and the manner in which it is burned. The dispersion and ground level concentrations of these emissions are determined by a complex interaction of the physical characteristics of the plant stack, physical chemical characteristics of the emissions, meteorological conditions at or near the site during the time the emissions travel from the stack to the ground level receptor, topographical conditions of the plant site and surrounding areas, and the nature of receptors (e.g., people, crops, and native vegetation).

4. The largest wastewater streams from thermoelectric plants are typically rather clean cooling water and can be either recycled or discharged to a surface water body with minimal effects on chemical quality. The impacts of waste heat on ambient water temperature need to be considered, however, when evaluating plants for which once-through cooling is being considered. A small ambient temperature increase can radically alter aquatic plant and animal communities. Other effluents from thermoelectric projects are less plentiful but can significantly affect water quality. For example, liquid effluents from coal-fired power plants include discharges from cooling system blowdown, boiler blowdown, demineralizer backwash and resin regenerator wastewater, ash transport wastewater, runoff from coal piles, ash piles and the site, as well as other miscellaneous low-volume wastewater and discharges from accidents or spills. Trace metals, acids, and other chemicals in various combinations are found in these effluents. Oil spills have a negative impact on water quality at oil-fired facilities.

5. Because a number of the impacts can be avoided altogether or mitigated more successfully and at less cost by prudent site selection, "Plant Siting and Industrial Estate Development" should be read in conjunction with this section. (For a summary of the potential environmental impacts that can occur during the construction and operation of thermoelectric plants, see Table 10.9 at the end of this section.)

Special Issues

Global and Transboundary Impacts

6. Emissions from thermoelectric projects can act as precursors of acid precipitation, particularly when coal with its high sulfur content is the fuel. Acid precipitation accelerates the deterioration of buildings and monuments, radically alters aquatic ecosystems of certain lakes, and damages vegetation in forest ecosystems. The combustion of fossil fuel in thermoelectric plants also generates CO_2 and NO_x, and global warming has been attributed to increases in CO_2 and NO_x in the atmosphere. However, it is currently impossible to predict the exact contribution of specific emissions from a particular thermoelectric project to these regional and global problems.

Cooling Water and Waste Heat

7. Many generating plants which use steam also have once-through cooling systems. If the high volume of water that large plants of this type require is taken from natural water bodies such as rivers and bays, there is a risk of mortality to aquatic organisms from entrainment and impingement in the cooling system. This can significantly reduce populations of fish and shellfish, some of which may be commercially important.

8. Heated water discharges can elevate ambient water temperatures. This can radically alter existing aquatic plant and animal communities, favoring organisms which are suited to higher temperatures. The new communities are then vulnerable to the opposite effect, namely, sharp reductions in ambient temperature following plant shut-downs because of failure or scheduled maintenance.

9. Use of evaporative cooling towers reduces the volume of water which must be withdrawn for cooling to that needed as make-up water to offset evaporation. Towers eliminate thermal discharge but produce cooling tower blowdown which must be discharged. In colder climates, there is another alternative: the temperature of cooling water discharge can be reduced by beneficial use of waste heat in the form of hot water or steam, e.g., for heating buildings or aquaculture ponds.

10. Either form of cooling entails some consumptive loss of water, thereby reducing the volume available for drinking, irrigation, navigation, and other uses in water short areas.

Impacts on the Community

11. One of the major impacts from power plants involves the influx of workers for building the plant. Up to several thousand workers may be required during the several years of construction of a large plant and up to several hundred workers for its operation. There is the potential for great stress where the host

community is small. A "boom town" condition or induced development can result. This can have a significant negative effect on the existing community infrastructure: school, police, fire protection, medical facilities, and so forth. Similarly, the influx of workers from other localities or regions will change local demographic patterns and disrupt local social and cultural values, as well as living patterns of the residents. Another potential impact is the displacement of the local population because of land requirements for the plant site and associated facilities. Significant disruption of local traffic can occur from the construction and operation of a thermoelectric plant. Finally, large power plants can be visually obtrusive and noisy.

Project Alternatives

12. The environmental assessment should include an analysis of reasonable alternatives to meet the ultimate objectives of the hydroelectric project. The analysis may lead to alternatives that are more sound from an environmental, sociocultural, and economic point of view than the originally proposed project. A number of alternatives need to be considered:

- no action (i.e., examine the consequences of taking no action to meet the expected demand needs)
- alternative fuels
- energy and load management alternatives
- site location alternatives
- alternative heat rejection systems
- alternative water supply/intake alternatives
- plant and sanitary waste discharge alternatives
- solid waste disposal alternatives
- engineering and pollution control equipment alternatives
- management control alternatives
- social structure alternatives including infrastructure and employment

13. Alternatives should be evaluated as part of the conceptual design process; however, those alternatives that provide cost-effective environmental control are preferred. The appropriateness of these alternatives should be addressed in relation to environmental and economic factors.

Management and Training

14. Because of the major environmental considerations involved in the construction and operation of a thermoelectric project, a team of environmental engineers and scientists need to be a part of the design and management staff for the facility. This group should work with the power plant engineers in all phases of the project that have environmental implications. Depending on the education and experience of the environmental staff, a training program in the environmental management of thermoelectric projects may be warranted.

A number of environmental discipline specialties that relate to management of thermoelectric projects need to be understood, including the following:

- ambient air quality monitoring, modeling, and pollution control
- water resources monitoring, modeling, and pollution abatement
- solid waste management and control and industrial hygiene
- toxic substances control and hazardous waste management
- noise abatement
- natural resource protection and land use planning
- socioeconomic impact assessment

15. Environmental training may be required for general impact assessment concepts, methodologies, monitoring theory and methods, data collection and analysis, and pollution control strategies. The training should be done as part of the environmental assessment phase of the project and with assistance from the environmental consultant. If at all possible, the environmental staff should be involved in the environmental assessment study. This will ensure an understanding of the environmental assessment of the project. In particular, staff workers must have an understanding of the rationale for the recommended mitigation and monitoring that they may be implementing. Training should be given to the technical staff and supervisory staff who will interface with the power plant engineers and managers.

16. Staff training in and management enforcement of standard operating and maintenance procedures, as well as health and safety procedures will be required to minimize environmental and health and safety impacts of the plant once it is in operation.

17. Often, there are no emission limitations or air quality standards in the borrower country that would affect potential thermoelectric projects. The World Bank has criteria that can serve as guidelines in lieu of country air quality standards. These environmental guidelines and other recognized criteria (e.g., studies of known effects) should be used as limits to protect human health and the environment.

18. Local, regional, and national environmental agencies involved in the review, approval, and oversight of the project may also need training to monitor and enforce compliance during the construction and operation of the project.

Monitoring

19. The purpose of a monitoring program is to provide information that the predicted impacts from a project are within the engineering and environmental acceptable limits, and to provide early warning information of unacceptable environmental conditions. Monitoring for thermoelectric projects should begin before construction to determine baseline conditions. Construction and operational monitoring will determine the degree and significance of impact that will occur during these phases of the project. Normally a year of preconstruction monitoring will be sufficient to characterize the environmental resources potentially affected by the project. The length of construction and operational monitoring will

depend on the environmental resource that is being affected and the expected duration of the impact. For example, if a continuous cooling water discharge is planned, then weekly or daily water quality monitoring may be needed for the life of the facility. Specific monitoring programs will be required depending on the type of thermoelectric project and the type of resources predicted to be affected.

20. Continuous air monitoring of the primary pollutants emitted from the facility will be required. Monitors should be established to measure emission concentrations and ground level concentrations at previously defined air quality receptor locations (e.g., residential areas, agricultural areas, etc.). Meteorological conditions for the site need to be characterized for air modeling purposes. If appropriate meteorological data are unavailable, then meteorological monitoring will be necessary.

21. Air monitoring of the workspace for dust, noise, and levels of toxic gases is necessary to protect operating personnel.

22. The type and nature of the wastewater discharge will determine if surface water quality monitoring will be required. Expected pollutants should be measured as well as water quality parameters that are important for human health and public welfare. If not more frequent, seasonal water quality monitoring may be necessary. Groundwater monitoring may be required if contamination of groundwater is predicted. Monitoring should be conducted upstream at the point of discharge, and downstream from the point of discharge in any receiving water body used by the public or considered environmentally significant (e.g., rivers, drinking and irrigation wells). Geophysical testing of the site may be required to characterize geological conditions under the proposed facility. If groundwater is proposed for cooling, then a pump test may be required to determine groundwater quantity and quality.

23. Biological monitoring may be appropriate if important biological resources occur near the project and are predicted to be affected, e.g., the discharge of cooling effluents into a estuary. In this case, sampling of representative kinds of aquatic organisms would be necessary. Important air quality receptors (e.g., sensitive crop species) and downwind of the stacks may require monitoring if adverse impacts are predicted. Sampling would be seasonal. Monitoring of the social environment may be warranted to ensure that infrastructure impacts are within acceptable limits.

24. The monitoring program should be designed to provide scientifically defensible information useful for determining the status of environmental resources affected by the thermoelectric project, to provide information to predict future effects, and to provide information for management decisions on possible mitigation if observed or predicted impacts are considered unacceptable.

Table 10.9. Thermoelectric Projects

Potential Negative Impacts	Mitigating Measures
Direct	
1. Air emission effects to human health, agriculture, and native wildlife and vegetation.	1. • Locate facility away from sensitive air quality receptors. • Design higher stacks to reduce ground level concentrations. • Use cleaner fuels (e.g., low sulfur coal). • Install air pollution control equipment.
2. Increased noise and vibration.	2. • Use lower rated equipment. • Control the timing of noise and vibration to least disruptive periods. • Install noise barriers.
3. Change in surface water and groundwater quality.	3. • Treat discharges chemically or mechanically on-site. • Prevent groundwater contamination through use of liners. • Use deep well injection below potable zones. • Construct liners for ponds and solid waste disposal areas. • Dilute effluent at point of discharge.
4. Toxic effects of chemical discharges and spills.	4. • Develop spill prevention plans. • Develop traps and containment systems and chemically treat discharges on-site.
5. Thermal shock to aquatic organisms.	5. • Use alternative heat dissipation design (e.g., closed cycle cooling). • Dilute thermal condition by discharging water into larger receiving water body. • Install mechanical diffusers. • Cool water on-site in holding pond prior to discharge. • Explore opportunities to use waste heat.

Table 10.9. Thermoelectric Projects (continued)

Potential Negative Impacts	Mitigating Measures
Direct (continued)	
6. Entrainment and impingement of aquatic organisms.	6. • Select water intake in area that avoids significant impact. • Install screens to eliminate entrainment and impingement.
7. Change in surface water and groundwater quantity.	7. • Develop water recycling plan.
8. Change in surface water flow and discharge.	8. • Construct drainage ways and holding ponds on-site.
9. Vegetation removal and habitat loss.	9. • Select alternative site or site layout to avoid loss of ecological resources. • Restore or create similar vegetation or habitats.
10. Dredging and filling of wetlands.	10. • Select alternative site or site layout to avoid loss of wetlands. • Restore or create similar wetlands.
11. Avian hazards from stacks, towers, and transmission lines.	11. • Site stacks and tower away from flyways. • Install deflectors, lights, and other visible features.
12. Human population displacement.	12. • Select alternative site or site layout to avoid displacement. • Involve affected parties in the resettlement planning and program. • Construct socially and culturally acceptable settlements/infrastructure development (see "Involuntary Resettlement" section).
13. Disruption of traffic.	13. • Develop traffic plan that includes phasing road use by workers. • Upgrade roads and intersections.

Table 10.9. Thermoelectric Projects (continued)

Potential Negative Impacts	Mitigating Measures
Direct (continued)	
14. Modification of historically or archaeologically significant structures or lands (e.g., churches, temples, mosques, cemeteries).	14. • Select alternative site or site layout. • Develop and implement "chance find" procedures to recover, relocate or restore structures (see "Cultural Property" section for detailed discussion). • Fence or construct other barriers to protect structures or lands.
15. Visual impact on historical, archaeological, and cultural resources and on landscapes.	15. • Select alternative site or site layout. • Construct visual buffers (e.g., plant trees).
16. Worker exposure to dust from ash and coal.	16. • Provide dust collector equipment. • Maintain dust levels ≤ 10 mg/m^3. • Monitor for free silica content. • Provide dust masks when levels are exceeded.
17. Worker exposure to toxic gases leaking from broilers.	17. • Maintain boilers properly. • Monitor concentrations with levels not to exceed: SO_2 5 ppm CO 50 ppm NO_2 5 ppm
18. Worker exposure to excessive noise.	18. • Maintain noise levels below 90 dBA, or provide ear protection.

Table 10.9. Thermoelectric Projects (continued)

Potential Negative Impacts	Mitigating Measures
Indirect	
1. Induced secondary development including increased demands on infrastructure.	1. • Provide infrastructure plan and financial support for increased demands. • Construct facilities to reduce demands.
2. Changes in demographic patterns and disruption of social and cultural values and patterns.	2. • Develop plan to educate workers on sensitive values and patterns. • Provide behavioral and/or psychological readjustment programs and services.

FINANCING NUCLEAR POWER: OPTIONS FOR THE BANK

1. This section reviews the Bank's practice with respect to financing nuclear power. Although the Bank has no formal and specific "Nuclear Energy Policy," this section is based on the Bank's experience to date, and on the best judgement available. It does not lay down Bank policy; rather it summarizes the options Task Managers may want to review if confronted with decisions concerning nuclear power. The second section outlines the current nuclear power situation. The third section reviews the major economic, financial, environmental, and political issues. The last section examines the major options for the Bank.

2. The Bank has never directly financed a nuclear power reactor. Until the early 1970's, the suppliers of nuclear plants were limited and credits were available from bilateral sources. The Bank took the position that, as the financier of last resort, it was unnecessary for its funds to be used for this purpose. In addition, given the limited number of suppliers at that time, procurement on the basis of International Competitive Bidding was not possible.

3. During this time, world-wide concerns about the costs and safety of these reactors had been increasing. This technology had not exported well. Costs typically had come in at two to three times the original estimates, delays had been substantial, and production problems had resulted in output well below capacity. It was a technology which, if used safely, would require rigorous standards of construction, maintenance, and operation -- areas in which developing countries have serious problems. Although some developing countries (Korea, Taiwan, Hungary, etc.) have done rather well in terms of construction period, capital costs and operations, others have done poorly because of technical, financial, and institutional reasons. In addition, the special characteristics of nuclear waste create problems which remain unresolved in many countries.

4. Military governments have been important proponents of acquiring nuclear technology with obvious implications for their country's future capacity to produce nuclear weapons. Many programs had close links to the military and access to information required to evaluate the technology was not available. This raised security concerns both among neighboring countries and within an increasingly vociferous international constituency. The opposition was strengthened by the decision of some countries not to adhere to the Nuclear Non-Proliferation Treaty, which represented an attempt by the world community to separate civilian and military uses.

5. Some countries did not rely on the military arguments in favor of nuclear power, but emphasized instead the links between this option for production and their domestic industry, with a view to specializing in what is seen as a modern high technology industry. While recognizing that it might not be the cheapest way of generating electricity at present, they assumed that it might have a competitive role at some future date. Thus, it was seen to be in the national interest to acquire nuclear experience early in order to have nuclear power as an option in the future.

Recent Developments

6. The Bank's position against financing nuclear energy requires re-examination in light of some major developments over the last decades. First, the nuclear industry has grown. There are now over 430 nuclear reactors producing electric power in 26 countries. The number of suppliers has increased and there is a large body of knowledge and experience in the construction, design and operation of these plants (over 4,000 reactor-years). Most developed countries generate modest portions of their electric power from nuclear reactors. Sixteen percent of the world's electric power production is nuclear, which is almost as large as hydro (18%).

7. Second, countries with manufacturers and capable of exporting nuclear reactors are rather few (the United States, Soviet Union, France, Canada, West Germany, Japan and, possibly, a few more for components and systems). It is doubtful that governments would be willing to subsidize exports of nuclear technology today, other than funding for research and development (R&D). As a result, the Bank might be asked to assist in putting together a financing package for a nuclear power plant. The request may be expressed in terms of assistance for a build-operate-transfer (BOT) or build-own-operate-transfer (BOOT) scheme which the Bank is promoting for other power projects.

8. Third, the increase in energy prices in the seventies made nuclear energy appear as a more attractive option, particularly when security of oil supplies also came to represent a major concern. Many developing countries looked on nuclear power as an increasingly attractive option for the long run and were prepared to pay the costs for acquiring the technology. The subsequent decline in oil prices reduced the attractiveness of nuclear power, although the crisis in the Gulf rekindled efforts by the industry to promote the nuclear option (see para 11 below).

9. Fourth, power systems in several developing countries are now large enough to utilize nuclear plants. For many years the minimum size nuclear plants were so large relative to the power systems of most developing countries that they were precluded for reasons of cost and system security. Standardized 600 MW plants are now available which could be considered for grids with a 4,000 MW installed capacity.

10. Fifth, the existence of the international Treaty on Nuclear Non-Proliferation, with its safeguards on the use and inspection of nuclear materials, has eased though not eliminated concerns about the diversion of materials for military purposes. The willingness of the USSR to take back and process its waste, and the more limited take-back agreement of France have also eased these concerns.

11. Sixth, growing concerns about the impacts of fossil fuel combustion on the earth's atmosphere have led the nuclear industry to promote nuclear power as a more benign alternative.

12. Seventh, the accidents at Three Mile Island (1979) and Chernobyl (1986), and the difficulties in safe disposal of radioactive wastes, have heightened public concern for the safety of the technology and have generated strong opposition to investments in nuclear plants. Bank lending for energy, even if the connection with nuclear power is indirect, will be subject to public scrutiny.

13. Eighth, Bank lending for a specific power project must take into account the cost structure of the sector as a whole. The growing number of nuclear plants is forcing the Bank to consider how these high-cost plants are to be treated in the financial and economic evaluation of power systems. The membership of Eastern European countries with the large nuclear power bases (e.g., Hungary, 39%) has increased the need to examine these issues.

Review of Major Issues

Economic and Financial Issues

14. The Bank justifies its loans for electric power investments by showing that they represent the least cost way of meeting a particular power need. Until recent years, the few nuclear plants that existed were not considered part of the power system because usually they were the assets of another agency with a multiplicity of objectives (research, technology development, etc.) among which producing power was merely a by-product. Power is sold to the system at rates set by the more conventional sources with the subsidy being borne by the nuclear agency, not the consumer of electric power.

15. The financial return on a power project and the price charged for its output is usually dependent on system wide tariffs and costs. Because the costs of nuclear generating plants are so large, they have significant impacts on total system costs and therefore can affect the return on any conventional investment in the system.

16. It is difficult to obtain accurate figures on the cost of nuclear reactors because of the specific circumstances surrounding their construction and use in both the developed and developing countries. This is compounded by the difficulties of comparing different types of stations, such as light water and heavy water, containment vessel and non-containment. The lowest developed country estimates are from France where published information suggests capacity costs of between US$1,500 and US$2,000 per kW. These costs may include capital and other subsidies that are difficult to quantify. In the United States, costs range between US$3,000 and $5,000 per kW. In Argentina and Brazil, costs are between US$5,000 and $8,000 per KW. Costs for conventional technologies range from US$500-600 per kW for combined cycle gas plants to $1,300-$1,600 for coal gasification, combined cycle plants, and slightly less for fluidized bed coal technology. Coal or oil steam turbines have costs of between US$800 and US$1300 per kW. Operating costs must be added to capital costs to obtain final electricity costs. Even with low operating costs, the high capital costs of nuclear plants preclude their being selected as the least cost alternative under any reasonable assumptions concerning prices of coal or oil.

17. Nuclear plants are thus uneconomic because at present and projected costs they are unlikely to be the least-cost alternative. There is also evidence that the cost figures usually cited by suppliers are substantially underestimated and often fail to take adequately into account waste disposal, decommissioning, and other environmental costs. Furthermore, the large size of many nuclear plants relative to developing country systems leads to risks of substantial excess capacity should demand fail to increase as predicted. A nuclear investment strategy lacks flexibility to adapt to changing circumstances. The high costs would require large increases in tariffs and could threaten the financial viability of the systems if nuclear power were a significant part of the total, though financial viability would not necessarily be threatened if its contribution is relatively small.

18. Nuclear plants also involve substantial financial and technical risks. Each plant represents an investment of US$1.5-2.0 billion. Failure to complete a plant on time involves added costs of between US$150 and US$200 million per year in financial charges. Delays of several years are not unusual and this, combined with the difficulties developing countries have in operating plants at rated capacity over prolonged periods, represents a substantial financial charge on most utilities.

Environmental Issues

19. The major environmental issue is whether nuclear plants (including the production of fuels, cooling systems, and waste disposal) can be operated within acceptable safety standards expressed mainly in terms of radioactive releases. There are major differences of opinion on what is acceptable in terms of both the costs and probabilities of accidents, particularly those of a catastrophic nature.

> **Catastrophic Failures:** Both nuclear and hydro plants have only a small probability of catastrophic failure, but some experts point to experience of systems failure in nuclear plants, where the exposure is much greater than in hydro dams (where the safety issue is a structural one). The worst case catastrophe for a nuclear plant is much worse than for a hydro plant because of the long-run health impacts (as at Chernobyl). In both cases, the consequences are borne by an involuntary population. To what extent a state is willing to expose its citizens to the risk of such events is a complex political value judgement. In the case of nuclear power, the lack of a meaningful historical record complicates these judgments.
>
> **Low-level Radiation:** The long-term effects of low-level radiation exposure are difficult to study because they are masked by other effects (chemicals, smoking, diet, etc.) and cannot therefore cannot be detected.

20. The environmental community is therefore strongly anti-nuclear. It emphasizes that the risk is one of involuntary exposure and that the environmental costs are high enough to rule out nuclear power even if it were otherwise economic. However, in recent discussions of the greenhouse effect, nuclear proponents have suggested that nuclear power might be part of the answer provided that safe, passive nuclear plants can be developed (see para 23). Some environmentalists are prepared to reserve judgement until such passively safe plants are proven.

21. Further complicating the issue is a perception of secrecy and lack of candor that characterizes the operation of nuclear power plants. In recent years, a number of accidents have raised doubts in the public mind as to the competence of the industry and the safety of the process. Many doubt the credibility of the industry.

22. The industry and supporting governments believe that if a plant is properly operated, the environmental costs are limited and the risks are acceptable. They claim that the safety record of the nuclear industry compares favorably with other energy sources -- even allowing for Chernobyl -- which was an inherently unstable technology not available to developing countries. The estimated loss of life from this disaster is less than from some large dam failures. However, the more permanent effects of Chernobyl, while still uncertain, are becoming better known and add substantially to the cost. In comparison, the Morvi dam accident in India in 1979 caused about 15,000 deaths and the Chernobyl numbers are climbing in that direction.

23. There has recently been increased concern about the impact of the release of CO_2 by fossil fuel combustion on the atmosphere -- the "greenhouse" effect. The nuclear industry has suggested that the advantage of nuclear power over fossil fuels in this respect could justify its higher costs. Opponents emphasize that conservation is a better alternative, particularly for developed countries, because it reduces the need for large new investments. The supporting evidence, particularly on the nuclear side, is controversial and incomplete.

24. The environmental issues have become highly emotional and politicized. Experts on both sides are suspect and differences are increasingly being resolved at great cost through the political and judicial processes as, for example, in the United States (Shoreham, Long Island), Sweden and Italy.

Political Issues

25. The decision to invest in nuclear power has become increasingly a political decision. The links between nuclear power and weapons development always have been prominent in the public debate, particularly where countries develop full fuel-cycle capabilities. The nuclear power industry denies this link, claiming that fuels produced from modern power reactors are not suitable for weapons production. A degree of linkage in the technologies cannot, however, be denied.

26. The Nuclear Non-Proliferation Treaty is intended to address this issue. The signatories agree not to transfer to and not to receive nuclear weapons from any party. States without nuclear weapons undertake not to manufacture or otherwise acquire them, and any assistance to this end is banned. A system of safeguards is administered by the International Atomic Energy Authority (IAEA) to verify compliance with the Treaty. In return for this obligation on the part of non-weapon states, the Treaty guarantees the "free transfer of nuclear technology for peaceful purposes without discrimination."

27. Some suppliers (e.g., Canada) insist that recipients of nuclear technology be signatories of the Treaty. Not all developing countries are signatories and some, notably Argentina and Brazil, have made a political issue of not signing, but this has not prevented them from obtaining the necessary technology.

28. A Bank position on financing nuclear power has to recognize that some of the major stockholders are exporters of the technology and in most cases are prepared to offer attractive terms to potential borrowers. In addition, the Soviet Union and other Council for Mutual Economic Assistance (CMEA) countries have shared production facilities for nuclear reactors and the industry is regarded as one of the more successful examples of integration in the CMEA market. (CMEA or Comecon is expected to be replaced in early 1991 by the Organization for International Economic Cooperation [OIEC]).

Options for the Bank

Basic Options

29. The Bank has two major policy options: either (a) the Bank is not prepared to finance any power system with nuclear capacity -- including conventional plants, transmission systems, and local distribution systems -- because of the high capital costs, the financial risks, the adverse environmental impact, and the possible military implications, or (b) the Bank is prepared to finance conventional components of a

system with some nuclear capacity, based on the usual Bank criteria -- economic efficiency (assessed after calculating full environmental costs), financial viability, non-availability of other financing, the use of international competitive bidding, etc. A third option would be to include under (b) direct investments in new nuclear plants, in existing nuclear plants, or in other components to support nuclear plants (e.g., transmission lines).

30. There are two crucial considerations -- economic and safety -- involved in any decision to finance nuclear plants directly. The economic case is clear: under present cost structures, the Bank would not finance new plants because they are uneconomic. In the unlikely event that nuclear plants become economic, the Bank would not finance them since there are other sources of funds available and, as financier of last resort, Bank funds are not required. In addition, of course, there is the concern of safety. The issue of safe construction and operation of a plant cannot be separated from its economic analysis, and the Bank would need to ensure that the institutional structure exists to support the safe operation of the plant. The Bank is not in a position to advise independently on the safety of nuclear plants. Nor can it supervise consultants who could advise on plant safety because nuclear power has strong proponents and opponents and it is difficult to find consultants with the objectivity the Bank requires.

31. The Bank should therefore express the view that although nuclear technology is uneconomical at present, there are great uncertainties about the future. It may be in the interest of a country to acquire the basic skills and experience to expand nuclear programs should it become necessary. This view has been accepted by the Bank, as in the case of the Chinese and Korean nuclear programs. However, it is a costly way to acquire nuclear technology and there is sufficient excess capacity in the exporting countries to permit the rapid transfer of the technology when and if it should become necessary. The Bank's support is not needed for these activities.

32. Where a nuclear plant is either already operating or under construction, the main task facing the Bank is to analyze the economic efficiency of the plant, ignoring sunk costs. The costs for operating plants, including environmental costs, would be compared with the costs of alternatives. For plants under construction, the costs of completion would also have to be taken into account. In addition, the Bank would require assurance that the capital costs of such plants are not incorporated in the utility rate base and that their power is sold at the cost of alternatives.

33. Given the fact (as discussed above) that the Bank does not have the capacity to independently judge the safety of nuclear plants, it is unlikely that such investments would be acceptable. It is true that we do not guarantee the safety of any project we finance. We rely on the appropriate national or international certification bodies. We require assurances that what we finance works in an acceptable manner. Normally, the Bank would rely on other agencies and authorities to play such a role, somewhat as we do in the case of dam safety.

34. This raises an important policy issue: to what extent can the Bank rely on others to play such a role for it when a plant has already been built? An important aspect of safety audits in nuclear plants relates to the demanding standards which have to be met during construction, as well as (more obviously) during operational phases, particularly during start up and shut down of the reactors. Once a plant has been operational, or even when it is partly constructed, such a safety audit cannot be performed in a way that would meet normal Bank criteria.

35. Similarly, although the regulatory body of an exporting country may be willing to certify the design of the plant, it cannot play a role in siting, construction, inspection, and other activities related to a plant in another country. Moreover, submission to safety inspections of all civilian installations related to the power sector would be necessary. Since the nuclear plant involves inflows and outflows of nuclear materials, all these flows would have to be accurately monitored and accounted for.

36. Bank lending for the energy sector requires a review of sector investments, institutions and policies. Nuclear plants in the power sector would not be economic; they are likely to be large "white elephants". All public investment programs include some mis-investments, however, and the Bank's position is based on pragmatic considerations regarding their size and degree of inefficiency. In this situation, the alternative facing the Bank is to reduce or even eliminate lending depending on the reasonableness of the sector investment plan as a whole.

37. A more difficult case concerns the Council for Mutual Economic Assistance countries in which nuclear capacity is a large proportion of their total generating capacity. In these, countries financial arrangements make the costs of nuclear plants difficult to determine. If the Bank plans to lend to the power sector in these countries, prior preparation of a position paper would be essential to explore the feasibility of carrying out an economic and environmental analysis of their nuclear investments, including their specific costs.

Summary

38. The Bank should continue to finance, through investment projects, specific non-nuclear components of an energy sector plan, provided the plan as a whole is reasonable. Reasonableness would be judged by criteria such as the relative size of any mis-investments (including nuclear energy), their rate of return, their impact on sector finances, etc. This is essentially what the Bank has been doing in the energy sector, as well as in other sectors. It would permit the Bank to continue to help improve policies and institutions in a sector which requires large amounts of capital. The Bank's effectiveness would depend on the size of its loans and its willingness to make a number of loans over a period of years. As noted in the preceding paragraph, the issue of money fungibility could still arise -- even for loans outside the energy sector -- but the degree of fungibility would be less than in a sector loan.

CEMENT

1. This category includes wet and dry process kiln facilities producing cement from limestone and lightweight aggregate kiln facilities producing aggregate from slate or shale. Rotary kilns are employed, operating at material temperatures of 1400° C. The principal raw materials are limestone, silica sand, clay, shale, marl and oxides of chalk. Silica, aluminum, and iron are added in the forms of sand, clay, bauxite, shale, iron ore, and blast furnace slag. Gypsum is added at the final phase of the process. All raw materials are received and stored in bulk. Cement kiln technology is employed worldwide. Cement plants are usually sited near limestone quarries in order to minimize raw material hauling costs. Whether or not they are juxtaposed, the environmental impacts of the quarrying operation should be considered in assessing the impacts of cement manufacturing (see the section on "Mining and Mineral Processing"). Figure 10.1 provides a generalized process diagram for typical wet and dry cement manufacturing processes.

Potential Environmental Impacts

2. Cement plants may have positive environmental impacts in waste management. The technology and the process are well-suited for the reuse or destruction of a variety of waste materials, including some hazardous wastes (see "Hazardous Materials Management" section). In addition, kiln dust not recycled at the plant can be used for liming soils, neutralizing acid mine drainage, and stabilizing hazardous waste or as a filler in asphalt.

3. Negative environmental impacts of cement operations are found in the following process areas: material handling and storage (particulates), grinding (particulates), and kiln and clinker cooling exhaust (particulates or "kiln dust", combustion gases containing carbon monoxide and dioxide, hydrocarbons, aldehydes, ketones, and oxides of sulfur and nitrogen). Water pollutants are found in spills of kiln feed (high pH, suspended solids, dissolved solids [chiefly potassium and sulfate]), and process cooling water (waste heat). Runoff and leaching from material storage and waste disposal areas can be a source of pollutants to surface water and groundwater. (See Table 10.10 at the end of this section for other examples of negative environmental impacts and recommended measures to avoid or mitigate them.)

4. Dust, especially free silica, constitutes a significant health risk to plant employees. Employee exposure to high noise levels is a hazard. Noise and truck traffic can be nuisances in the surrounding community.

5. Because a number of the impacts described can be avoided altogether or mitigated more successfully and at less cost by prudent plant site selection, "Plant Siting and Industrial Estate Development" should be read in conjunction with this section.

Special Issues

Particulate Emissions to the Atmosphere

6. Cement manufacturing involves hauling dusty or pulverized materials from quarrying the limestone to loading the finished product for shipment. Particulates are the single most significant cause of negative

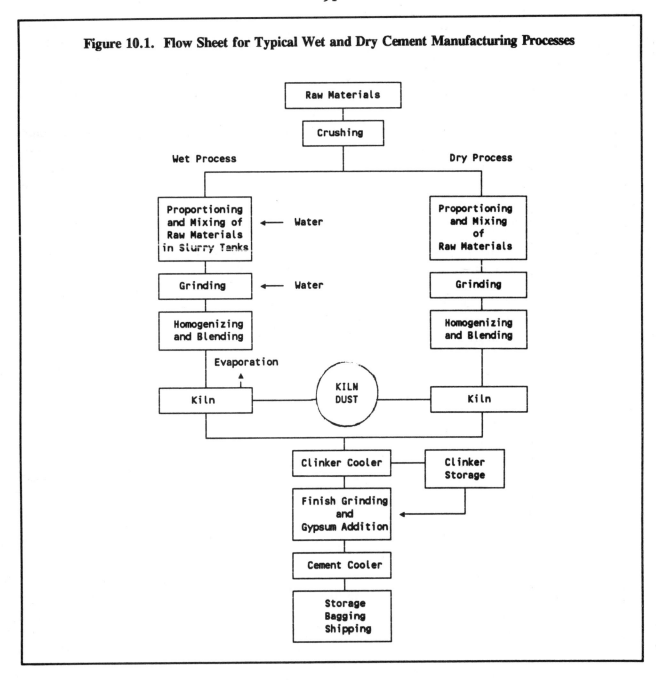

Figure 10.1. Flow Sheet for Typical Wet and Dry Cement Manufacturing Processes

environmental impact. Electrostatic precipitators or bag filters are a routine requirement to control particulate emissions from kilns. Control of dust resulting from hauling materials presents a more difficult challenge; conveyors, storage piles, and plant roads can be a greater cause of air quality degradation than mill and kiln exhausts. Mechanical dust collectors should be employed where practical, as on crushers, conveyors, and loading facilities. In most cases, collected dust can be recycled, reducing cost and minimizing solid waste production. Plant roads should be kept clean by using vacuum sweepers and/or sprinklers to prevent airborn dust from traffic and wind. Storage piles should be covered as much as possible. Trucks transporting materials to and from the plant should have tarpaulin covers and be limited in speed.

Liquid Waste Discharges

7. In so-called "dry" process plants, dry raw materials are fed to the kiln. The only effluent is cooling water which can be eliminated through use of cooling towers or ponds. In the "wet" process, raw materials are fed as a slurry to the kiln. In a few cases, plants may leach collected kiln dust to remove soluble alkali before refeeding it to the kiln. In such plants, clarifier overflow from the leaching process is the most severe source of water pollution; it requires neutralization (possibly by carbonation) before discharge.

Using Cement Kilns for Waste Recycling or Disposal

8. Waste oils, solvents, paint residues, and other combustible waste products have been used as supplemental fuels in cement kilns. The practice began in the United States, as early as 1979, for energy conservation and fuel cost reduction and has been satisfactory in terms of both product quality and environmental impact. Some solid wastes can also serve as fuel, such as the use of discarded tires. Raw material requirements for can be partially satisfied by waste products (and routinely used) from other industries: gypsum from phosphoric acid plants, roasted pyrites from sulfuric acid production, blast furnace slag, and fly ash from coal-fired generating stations.

9. High flame temperatures and the nature of the product make cement kilns attractive for the destruction of a variety of hazardous organic materials. With proper management, kilns constitute a much less costly alternative to dedicated waste incinerators. Tests by USEPA and others have demonstrated that destruction of organic compounds including, PCBs and organochlorine and organophosphorus pesticides, equals or surpasses that achieved in hazardous waste incinerators that operate at lower temperatures. Many toxic metallic compounds can also be burned in cement kilns in quantities that are small enough not to adversely affect product quality or safety, since they are bonded to the clinker and become part of the product. Lead, however, requires special attention; as much as one-half of the amount introduced leaves the kiln and precipitates with kiln dust. Dust recycling builds up the lead concentration to the point where lead also bonds to clinkers, but a small amount (0.2 to 1.0 percent) will escape in flue gases. Thallium will be emitted in kiln flue gases -- that is, it does not bond to the solids -- and studies of the behavior of mercury have been inconclusive.

10. Use of cement kilns for hazardous waste disposal requires special provisions in plant operating procedures, staffing and monitoring for worker protection, public health, and environmental quality. It also requires the development of emergency response plans and the involvement of the potentially affected community. For more details, the following sections should be consulted along with the discussions on Management and Training, and Monitoring in this section: "Industrial Hazard Management", "Hazardous Materials Management", and "Plant Siting and Industrial Estate Development."

Project Alternatives

Site Selection

11. General issues to consider in industrial plant siting are discussed in the section on "Plant Siting and Industrial Estate Development." The nature of cement production is such that impacts on air quality and

impacts of raw material extraction and transportation of bulk materials to and from the plant warrant special attention in evaluating alternative sites. Regions with substandard ambient air quality or populated localities with meteorological or topographic characteristics leading to limited air circulation are inappropriate. If the plant's demand for raw materials will necessitate opening additional quarry sites, they should be identified (if known) and their environmental impacts considered as part of the project. Proximity to sources of waste products that can serve as fuel and raw material substitutes or supplements is a positive factor in site selection. All other things being equal, plant location near the limestone source is desirable in order to minimize transportation costs (1 ton of cement requires 1.3-1.4 tons of limestone).

Alternative Fuels

12. Cement kilns can be fired with either coal, oil or gas, or a combination of them. Waste material can be used as supplementary fuel. These decisions also have implications for environmental quality and for the magnitude of pollution control investment.

    (a) **Air Pollution Control**

        alternatives for capture of kiln dust:

- electrostatic precipitator
- baghouse filter

        alternatives for capture of dust from clinker cooler:

- granular bed filter
- electrostatic precipitator
- baghouse filter

        alternatives to control dust from other operations:

- cover or enclose conveyors, crushers, material transfer points, storage areas
- install mechanical dust collectors and/or baghouse filters where needed
- pave plant roads
- vacuum sweepers for plant roads
- sprinklers for plant roads and storage piles
- latex stabilizing sprays for storage piles

    (b) **Water Pollution Control**

- recycling wet process waste water to kiln
- cooling towers and ponds
- runoff control from waste or raw material piles by diking
- infiltration control from waste or raw material piles by lining

Management and Training

13. The potential negative impacts on air quality from all cement processes and on water quality from leaching plants necessitate institutional support for efficient conduct and supervision of pollution control and waste reduction. The plant staff should include a plant engineer trained in air and water pollution control and monitoring technologies being used. Manufacturers will frequently supply the necessary equipment operations and maintenance training, if requested. Standard operating procedures should be established for the plant and enforced by management. They should include pollution control equipment operation, air and water quality monitoring requirements, plant road and storage area cleaning, procedures to minimize adverse impacts during kiln start-up (when precipitators will not operate effectively), and instructions for notification and shutdown or other responses to pollution equipment failure.

14. Plant health and safety rules should be established, including procedures to keep exposure to nuisance dust and free silica below national standards (or where none exist, below World Bank limits), a program of routine medical examinations, and ongoing training in plant health and safety and good environmental housekeeping practices. (For further discussion, see the World Bank's Occupational Health and Safety Guidelines.)

15. If the plant is to be used for destruction of hazardous wastes, special procedures for handling the material on-site and responding to emergencies are necessary. The hazardous waste part of the operation should be supervised and carried out by specially-trained employees. Transportation and storage of materials should be carefully controlled by appropriate regulatory and public safety officials, and conducted in accordance with accepted practices regarding materials handling, notification, and emergency response. (See the "Industrial Hazard Management" section.)

16. Emission and effluent standards should be set for the plant, based on national regulations where they exist, or on World Bank guidelines where they do not. Government agencies charged with monitoring the operation of pollution control equipment, enforcing standards, and overseeing any associated hazardous waste destruction activities may require specialized training and should be provided with the necessary equipment and authority. The environmental assessment should include an evaluation of local capabilities in these areas and recommend appropriate elements of assistance to be included in the project.

Monitoring

17. Monitoring plans are necessary for plant- and site-specific. In general, however, monitoring at a cement plant should include the following: continuous flue gas opacity, periodic stack testing for particulates to calibrate and verify opacity monitors, kiln dust, flue gas and cement for any hazardous materials being burned, liquid waste streams for pH (continuous), total dissolved and total suspended solids, alkalinity, potassium and sulfate, working areas for fugitive dust, free silica and noise, receiving waters for pH, total suspended solids and ambient air quality for suspended particulates, storage piles for runoff and leaching, and inspection for adherence to safety and pollution control procedures.

Table 10.10. Cement

Potential Negative Impacts	Mitigating Measures
Direct: Site Selection	
1. Siting of plant on/near sensitive habitats such as mangroves, estuaries, wetlands, coral reefs.	1. • Locate plant in industrially zoned area, if possible, to minimize or concentrate the stress on local environmental services and to facilitate the monitoring of discharges. • Integrate site selection process with natural resource agencies to review alternatives.
2. Siting along water courses causing their eventual degradation.	2. • Site selection process should examine alternatives that minimize environmental effects and not preclude beneficial use of the water body. • Plants with liquid discharges should only be located on a watercourse having adequate waste-absorbing capacity.
3. Siting can cause serious air pollution problems for local area.	3. • Locate in an area not subject to air inversions or trapping of pollutants, and where prevailing winds are towards relatively unpopulated areas.

4. Siting can aggravate solid waste problems in an area.	4. • Site selection should evaluate the location according to the following guidelines:
	  • plot size sufficient to landfill or dispose on-site
	  • proximity to suitable disposal site
	  • convenient for public/private contractors to collect and haul solid wastes for final disposal

Direct: Plant Operation	
5. • Water pollution from discharge of liquid effluents and process cooling water or runoff from waste piles.	5. • Laboratory analysis of liquid effluent should include TDS, TSS, salts, alkalinity, potassium, sulfates, and in-situ pH temperature monitoring.

Table 10.10. Cement (continued)

Potential Negative Impacts	Mitigating Measures
Direct: Plant Operation (continued) • Plant: Total Suspended Solids (TSS), Total Dissolved Solids (TDS), temperature, pH • Materials storage piles runoff: TSS, pH	**All Plants** • No cooling water discharge. If recycling not feasible, discharge cooling water provided receiving water temperature does not rise >3°C. • No discharge of slurry tank wash or spills. • Maintain pH level of effluent discharge between 6.0 and 9.0. **Non-Leaching Plants** • TSS <5 g/ton product • TDS no greater than levels of water incoming to plant **Leaching Plants** • TSS <150 g/ton product • TDS <1.5 kg/ton product **Material Storage Piles** • Minimize rainfall allowed to percolate through piles and runoff in uncontrolled fashion. • Line storage areas. **Equipment Washing, Road Washing, Other Washing** • <150 g/ton product during equipment cleaning operations or during periods of rainfall. • Plant housekeeping procedures must reflect desired level of mitigation.

Table 10.10. Cement (continued)

Potential Negative Impacts	Mitigating Measures

Direct: Plant Operation (continued)

6. Particulate emissions to the atmosphere from all plant operations, crushing, material handling, kilns, clinker coolers.

    6. • Control particulates by fabric filter collectors.
    • Control kiln particulate emissions by electrostatic precipitator dust collectors, with moisture conditioning required for dry process operations.
    • Control particulates on dry basis as follows:
        • from kiln, 150 g/ton feed
        • from clinker cooler, 50 g/ton feed
        • ground level outside plant fence, 80 $\mu g/m^3$
        • stack discharge, 100 $\mu g/m^3$

7. Particulate emission from ground sources (fugitive dust particulates), roads, piles.

    7. • Control measures include:
        • road treatment
        • water spray on piles
        • use of industrial vacuum cleaner
        • limit speed to 20 km/hr

8. Kiln gaseous emission of SO_x to the atmosphere from fuel burning.

    8. • Control by natural scrubbing action of alkaline materials and enhanced by utilization of preheater kilns, and use of exhaust gases to dry raw materials in grinding.
    • An analysis of raw materials during feasibility stage of project can determine levels of sulfur to properly design emission control equipment.

    <u>Inside plant fence</u>
    • Annual arithmetic mean: 100 $\mu g/m^3$
    • Maximum 24 hour peak: 1000 $\mu g/m^3$

Table 10.10. Cement (continued)

Potential Negative Impacts	Mitigating Measures
Direct: Plant Operation (continued)	**Outside plant fence**
	• Annual arithmetic mean: 100 $\mu g/m^3$ • Maximum 24 hour peak: 500 $\mu g/m^3$
9. Kiln gaseous emissions of NO_x to the atmosphere from fuel burning.	9. • Reduce NO_x by use of coal fuel and preheater/precalciner kilns. • Use of vegetative material or chemical wastes from other local industries should be carefully reviewed since these fuels may increase NO_x releases to the atmosphere.
10. Air pollution during start-up of the kiln (and the electrostatic precipitator is not available).	10. Where possible, conduct start-up when wind direction is not directed to ecologically sensitive or populated areas.
11. Air pollution as a result of electrostatic malfunction.	11. • Design precipitator with parallel chambers to enable the use of one part of the precipitator when the other is under repair. • Enforce kiln shut-down when precipitator is completely out-of-order.
12. • Burning hazardous wastes or waste oils as supplemental fuels could emit toxic air pollutants as products of incomplete combustion and metals such as lead to the atmosphere. • Handling and storage of hazardous wastes pose risks to community and environment.	12. • Although studies have shown that most organic materials are destroyed at an efficiency of 99.99 percent and metals are adsorbed to cement dust that is collected by the air pollution control system. • Care must be exercised to ensure that (a) hazardous waste and waste oils are analyzed before approval for burning, and (b) kiln operating efficiency is maintained.

Table 10.10. Cement (continued)

Potential Negative Impacts	Mitigating Measures
Direct: Plant Operation (continued)	
	• Add the waste at the "hot" end of the kiln.
	• Develop hazardous waste handling procedures and contingency plans (see "Hazardous Materials Management" section).
13. Surface runoff of constituents leached from kiln dust, raw materials, clinker, coal and other substances frequently stored in piles on the facility grounds can pollute surface waters or percolate to ground waters.	13. • Rainwater percolation and runoff from solid materials, fuel and waste piles can be controlled by covering and/or containment to prevent percolation and runoff to ground and surface waters.
	• Diked areas should be of sufficient size to contain an average 24 hour rainfall.
Indirect	
14. • Occupational health effects on workers due to fugitive dust, materials handling or other process operations.	14. • Facility should implement a Safety and Health Program designed to:
• Accidents occur at higher than normal frequency because of level of skill or labor.	• identify, evaluate, monitor, and control safety and health hazards at a specific level of detail
	• address the hazards to worker health and safety
	• propose procedures for employee protection
	• provide safety training
15. Regional solid waste problem exacerbated by inadequate on-site storage.	15. • Plan for adequate on-site disposal areas or use of kiln dust or other by-products as local fill material, assuming screening for hazardous characteristics of the leachate is known.
	• Use kiln dust for soil liming, neutralizing acid or stabilizing hazardous waste.

Table 10.10. Cement (continued)

Potential Negative Impacts	Mitigating Measures
Indirect (continued)	
16. Transit patterns disrupted, noise and congestion created, and pedestrian hazards aggravated by heavy trucks transporting raw materials, fuel or cement to/from facility.	16. • Site selection can mitigate some of these problems. • Special transportation sector studies should be prepared during project feasibility to select best routes to reduce impacts. • Transporter regulation and development of emergency contingency plans to minimize risk of accidents during transport of waste fuels.
17. Mining of limestone locally to provision cement facility can create conflicts with other industries, such as housing and construction, that rely on some similar resources and can aggravate erosion/sedimentation of water courses by uncontrolled or unrestricted operations.	17. • Plan for limestone resource usage to fit availability and impose restrictions on manner of quarrying. • Coordination with responsible agency-in-charge to examine site reclamation options once facility is de-commissioned. • Provide plan for limestone mine restoration.

CHEMICAL AND PETROCHEMICAL

1. The chemical and petrochemical industry sector comprises a multitude of processes and is by far the most diverse category of industries. It can be subdivided into the following groups: (a) inorganic chemicals, (b) organic chemicals, (c) petrochemicals and (d) fine chemicals, pharmaceuticals, synthetic dyes and explosives.

The fertilizer industry, even though a part of the chemical and petrochemical industry, is dealt with separately in this section.

2. The inorganic chemicals group includes the manufacture of chlorine/alkali, calcium carbide, inorganic acids, salts, phosphor and phosphor compounds, hydrogen peroxide, inorganic pigments (e.g., titanium dioxide), and many metal salts from the acids mentioned. Inorganic chemicals such as ammonia, nitric acid, urea, phosphoric acid, etc., are discussed in the section on fertilizer manufacturing.

3. Petrochemicals form a separate category of organic chemicals. Most of these chemicals are based on petroleum, natural gas or coal as raw material and many are produced in big quantities (between 1,000 ton/year for specialty products and 500,000 ton/year manufacturing units for the commodities).

4. Many of the petrochemicals require liquid or gas storage facilities. Examples are ethylene, methanol, ethanol, acetic acid, acetone, adipic acid, aniline, benzene, caprolactam, compounds from chlorine and fluorine with aliphatic or aromatic chemicals, dinitro- and trinitro-toluene, formaldehyde, and alcohols. Solid products include synthetic resins, plastics and elastomers, rubber, melamine, nylon, polyester, polyolefins and polyvinylchlorides. Other products like cellulose and sugar-based chemicals, although not petrochemicals, can be accommodated in this group.

5. The fine chemicals and pharmaceuticals form a separate group mainly because of the different industrial approach. These chemicals are nearly always manufactured in small quantities from either petrochemicals, natural products or inorganic chemicals. This group comprises all synthetic fragrance and flavor compounds, synthetic dyes, pharmaceutical intermediates and end products.

6. Modern chemical manufacturing facilities generally entail the construction of separate wastewater treatment plants to allow reuse of water after the pollutants have been removed by chemical or physical methods to an extent that their concentration is reduced to a level that is considered tolerable. Preferably, raw material and product storage facilities should be designed and built with containment provisions like double walled tanks, diking, or concrete walls and tank leak detection systems.

Potential Environmental Impacts

7. Most of the materials used in the manufacture of chemicals and petrochemicals are flammable and explosive. Although many chemicals and petrochemicals are toxic, some are also carcinogenic. Potential explosion hazards are much more severe compared to, for example, the refining industry because of highly reactive compounds and the high pressures involved in manufacturing and handling.

8. Highly toxic materials that cause immediate injury, such as phosgene or chlorine would be classified as a safety hazard. Others have long-term effects sometimes in very low concentrations. In studies on the production of chemicals and their environmental impact, toxicity, hazards, and operability considerations were found to play an important role. The potential wastes and emissions are a function of the types of compounds manufactured and the great diversity of processes and chemicals used in manufacturing them.

9. The negative environmental impact of production of chemicals can be severe. For information on chemical and health hazards, the National Institute for Occupational Safety and Health (NIOSH), a branch of the U.S. Department of Health and Human Resources (HHS), has published a guide book. The Dow and Fire and Explosion Index, published by the American Institute of Chemical Engineers (AICE), is used for information on fire and explosion hazards. (See also Table 10.11 at the end of this section for a summary of the negative environmental impacts from chemical and petrochemical production.)

10. Large quantities of water are used in the chemical industry for process, cooling and washing. During chemical manufacturing, water often becomes contaminated with chemicals or byproducts. The U.S. Environmental Protection Agency (EPA) has published a list of compounds from which water effluent guidelines have been established. Pollutants which may present a hazard if released into waterways and underground aquifers include toxic priority pollutants, carcinogenic compounds, suspended solids, and substances with high biochemical oxygen demand (BOD) and chemical oxygen demand (COD).

11. Groundwater and surface water resources can be negatively impacted by rainwater from tank farms, product discharge and processing areas, pipe tracks, cooling water blowdown, flushing and cleaning water and accidental release of raw materials and finished products. Runoff control measures, such as stormwater detention basins with treatment prior to discharge, are normally necessary to avoid such adverse water impacts.

12. Depending on the process used, air pollutants include particulate matter and a great number of gaseous compounds including sulfur oxides, carbon oxides and nitrogen oxides from boiler flues and process furnaces, ammonia, nitrogen compounds, and chlorinated compounds. These emissions result from several sources including process equipment, storage facilities, pumps, valves, vents and leaking seals.

13. Air emissions are controlled by use of incineration (stack flares), adsorption, gas scrubbing, and other absorption processes. Air quality standards to regulate emissions from chemical manufacturing facilities have been developed by the U.S. Environmental Protection Agency.

14. Solid wastes from the chemical process industry may include residuals from raw materials, waste polymers, sludges from boiler feed, tank cleaning or pollution control equipment, and ash from coal boiler operations. Waste material may be contaminated with chemical substances from the processes. The disposal of spent catalysts can generate an environmental problem in petrochemical industries. These days, most catalyst suppliers offer services for the return of spent catalysts.

Special Issues

Hazardous Materials Management

15. In some instances, wastes may represent a biological or radiation hazard. For instance, bio-industrial and pharmaceutical waste can contain undesirable microorganisms and viruses and radioactive materials through improper disposal. The following practices should be instituted when dealing with disposal of this type of solid waste.

- For hazardous or radioactive materials, there should be adequate treatment, storage and disposal (TSD) facilities.

- The borrowing country should have developed (or adopted from advanced foreign sources) and implemented regulations and standards governing the operation of those facilities and be able to monitor regulatory compliance.

- Laboratory and other support facilities should exist to provide adequate collection and analysis of environmental samples.

16. Special problems are posed by the production of explosive materials or of highly reactive chemicals. Here design considerations like ruptured discs, explosion, and fire walls have to be incorporated to minimize environmental and health risks at the work site and outside.

17. Special environmental problems are often generated by formulation plants where chemicals are mixed according to special formulations to serve the market. Examples are pesticide formulation plants, solvent formulation facilities, and explosives facilities. Environmental, health, and hazard procedures for these type of plants should be the same as for the chemical plants manufacturing the components that are blended. (For further discussion, see the section on "Hazardous Materials Management".)

Wastewater Minimization

18. There are two types of in-plant measures that can greatly reduce the volume of wastewater effluent. The first is to reuse water from one process in another; for example, to use blowdown from high-pressure boilers as feedwater for low pressure boilers or treated effluent as make-up water wherever possible. The second approach is to design systems that recycle water repeatedly for the same purpose. Examples are employing cooling towers or using steam condensate as boiler feedwater.

19. Good housekeeping, combined with good work practices, will further reduce waste flows. Examples are minimizing waste when sampling product lines, using vacuum trucks or dry cleaning methods for spills, applying sound inspection and maintenance practices to minimize leakage, and segregating waste streams having special characteristics for disposal (e.g., spent cleaning solution).

Noise

20. Chemical and petrochemical manufacturing facilities can result in significant noise emissions. Sources of noise include high speed centrifugal compressors, rotary screw compressors, control valves, piping systems, gas turbines, pumps, furnaces, flares, air cooled heat exchangers, cooling towers and vents. Typical noise levels range from 60 to 110 dB at a distance of one meter from the source. Although acoustical insulation is often the most practical solution, equipment manufacturers have sometimes low noise equipment lines available. The American Petroleum Institute has published guidelines on noise and noise control. The Construction Specifications Institute (CSI) provides guidelines on acoustical insulation specifications.

Project Alternatives

Site

21. General issues for consideration in industrial plant site selection are discussed in the section on "Plant Siting and Industrial Estate Development." The nature of the chemical manufacturing industry is such that the impacts of production, storage, and transportation on the environment warrant special attention in evaluating alternative sites. Besides emission and effluent considerations, another aspect that needs attention is the transport of raw materials to the site and of end products from the site. Often toxic or highly inflammable materials are involved, especially in the petrochemical industries that can pose special transportation problems. Emissions can have a negative impact on the surrounding ecology or on nearby living areas, such as villages or towns. Transportation through densely populated areas should be avoided.

Manufacturing Processes

22. Chemical manufacturing uses a wide variety of processing and storage equipment. In the design phase, special attention should be paid to alternative processes. An example is the choice of process for a chlorine/alkali electrolysis plant. Older designs are based on mercury electrolysis cells that pose a great environmental threat because of the mercury content of wastewater. Alternative processes are now available like the diaphragm (the presence of asbestos in the cells is a lesser hazard) and the membrane processes that do not use mercury.

Pollution Control

23. Air pollution and effluent control equipment is now available for practically every waste stream, gaseous or liquid. Air pollution control equipment includes gas scrubbing, membrane separation, cyclones, electrostatic precipitators, baghouse filters, catalytic reduction or oxidation, incineration, and absorption systems.

24. Wastewater effluent can be controlled through neutralization, evaporation, aeration, stripping, flotation, filtration, oil separation, carbon absorption, ion exchange, reverse osmosis, biological treatment, and land application of process wastewater.

Management and Training

25. The potential impacts of chemical and petrochemical manufacturing processes on air, water, and soil necessitate institutional support for efficient conduct and supervision of materials handling, as well as for pollution control and waste reduction. Facility personnel should be trained in air and water pollution control technologies being employed. Equipment manufacturers will frequently supply the necessary training in equipment operations and maintenance. Standard operating procedures should be established for the plant and enforced by management. They should include pollution control equipment operation, air and water quality monitoring requirements, instructions for operators to prevent malodorous emissions, and directives for notification of proper authorities in the event of accidental release of pollutants. Toxic and hazardous substance handling and management should be improved by detectors, alarms, etc., and special training of operating personnel.

26. Emergency procedures are necessary to provide for rapid and effective action in the event of accidents, such as major spills, fires, and/or explosions, that pose serious threats to the environment or to the surrounding community. Local government officials and agencies, as well as local community services (medical, firefighting, etc.), usually play key roles in emergencies of this sort and should be included in the planning process. Periodic drills are important components of response plans. (See the "Industrial Hazard Management" section for detailed discussion.)

27. Plant health and safety rules should be established and enforced. They should include:

- Provisions to prevent and respond to accidental gaseous releases or liquid spills of chemicals.

- Procedures to keep exposure to chemical vapors below accepted standards (see The National Institute for Occupational Safety and Health's Guide to Chemical Hazards).

- A program of routine medical examinations, if hazardous chemicals are handled, stored, processed or transported.

- Ongoing training in plant health and safety and in good environmental housekeeping practices.

- Emergency procedures (and regular drills) to provide a plan of action in case of a major spill, leak, explosion or fire.

(See the World Bank's Occupational Health and Safety Guidelines and the following sections in this chapter for further discussion: "Industrial Hazard Management", "Hazardous Materials Management", and "Plant Siting and Industrial Estate Development.")

28. Emission and effluent standards should be set for the plant based on national regulations where they exist or on World Bank guidelines where they do not. Government agencies charged with monitoring air and water quality, pollution control equipment operation, enforcing standards, and overseeing waste disposal activities should have the necessary equipment, authority, and required training to do so. The environmental assessment should include an evaluation of local capabilities in these areas and recommend appropriate elements of assistance to be included in the project.

Monitoring

29. Because of the great variety of chemicals and processes used, it is impossible to give a listing of all the chemicals that should be monitored. Continuous records on environmental monitoring should be maintained, and periodic reviews and corrective actions taken. Although monitoring plans are necessary for process/plant/site-specific, the following procedures should also be established:

- Continuous monitoring of combustion gases in boilers and furnaces for carbon monoxide and excess air and opacity.

- Periodic, or where critical, continuous monitoring of gaseous and particulate emissions for chemicals that are used or generated in the process. (For petrochemical plants these are mainly hydrocarbons, chlorine [containing compounds], hydrogen, oxygenated organic compounds, nitrogen and sulfur containing compounds.)

- Periodic, or where critical, continuous monitoring of all wastewater streams including used cooling water for compounds mentioned in the above section.

- Measurement of selected process parameters to monitor adequate operation of the pollution control equipment (such as flue gas temperatures to check the operation of scrubbers).

- Workspace air quality monitoring for all compounds used in the process. (Measurement of several often can be combined conveniently, e.g., the level of all organic compounds or of certain groups of compounds such as those containing chlorine.)

- Monitoring of ambient air quality in vicinity of plants for applicable pollutants, especially toxic and hazardous chemicals, through remote sensors and alarms.

- Measurement of stormwater discharges from plants and storage areas for applicable pollutants, pH and total suspended solids.

- Monitoring of receiving water quality downstream for dissolved oxygen and applicable pollutants.

- Periodic monitoring of groundwater quality to detect contamination from the process or storage area.

- Monitoring effects of solid waste management practices on ground and surface water resources.

- Monitoring all working areas of plants for ambient noise levels.

- Inspection for adherence to safety and pollution control procedures, timely revisions, and updating of safety and emergency plans.

- Testing of receiving water for pH, total suspended solids, and ambient air for particulate matters.

Table 10.11. Chemical and Petrochemical

Potential Negative Impacts	Mitigating Measures
Direct: Site Selection	
1. Siting of plant on/near sensitive habitats such as mangroves, estuaries, wetlands, coral reefs.	1. • Locate plants in industrially zoned area, if possible, to minimize or concentrate the stress on local environmental services and to facilitate the monitoring of discharges. • Involve natural resource agencies in site selection process to review alternatives.
2. Siting along water courses causing their eventual degradation.	2. • Site selection process should examine alternatives that minimize environmental effects and do not preclude beneficial use of the water body. • Plants with liquid discharges should be located only on a watercourse having adequate capacity to assimilate wastes in treated effluent.
3. Siting can cause serious air pollution problems for local area.	3. • Locate plants in an area not subject to air inversions or trapping of pollutants, and where prevailing winds are towards relatively unpopulated areas.

4. Siting can aggravate solid waste problems in an area.	4. • Site selection should evaluate the location according to the following guidelines:
	  • plot size sufficient for landfill or disposal on-site
	  • proximity to suitable disposal site
	  • convenient for public/private contractors to collect and haul solid wastes for final disposal

Table 10.11. Chemical and Petrochemical (continued)

Potential Negative Impacts	Mitigating Measures
Direct: Plant Operation	
5. • Water pollution from discharge of liquid effluents and process cooling water or runoff from waste piles. • Depending on the process, runs at too high TOS, BOD, COD, and pH.	5. • Laboratory analysis of liquid effluent should include applicable chemicals (depending on the process), TOS, BOD, COD, pH and in-situ temperature monitoring. **All Plants** • No cooling water discharge. If recycling not feasible, discharge cooling water provided receiving water temperature does not rise >3°C. • Maintain pH level of effluent discharge between 6.0 and 9.0. • Control effluent to specified limitations in Bank or other guidelines for specific process. **Processing, Storage and Dispatch Area** • Minimize rainfall allowed to percolate through piles and runoff in uncontrolled fashion. • Line open storage areas to collect all stormwater.
6. Particulate emissions to the atmosphere from all plant operations.	6. Control particulates by scrubbers, fabric filter collectors or electrostatic precipitators.
7. Gaseous emission of SO_x, NO_x, and CO and other applicable chemicals to the atmosphere from chemical processes.	7. Control by scrubbing with water or alkaline solutions, incineration, or absorption by other catalytic processes.
8. Accidental release of potentially hazardous solvents, acidic and alkaline materials.	8. • Maintain storage and disposal areas to prevent accidental release. • Provide spill mitigation equipment. • Provide area diking or double wall tanks.

Table 10.11. Chemical and Petrochemical (continued)

Potential Negative Impacts	Mitigating Measures
Direct: Plant Operation (continued)	
9. Accidental radiation/biological hazard release (pharmaceuticals).	9. Maintain certified storage and disposal facilities to minimize potential for release.
10. Noise.	10. Reduce noise impact by enclosing and insulating noise emitting processes or equipment in buildings or by use of other noise abatement procedures.
11. Surface runoff of chemicals, raw materials, intermediates, end products, and solid wastes frequently stored in piles on the facility site can pollute surface waters or percolate to groundwaters.	11. Rainwater percolation and runoff from solid materials, fuel and waste piles can be controlled by covering and/or containment to prevent percolation and runoff to ground and surface waters.
	• Diked areas should be of sufficient size to contain an average 24 hour rainfall.
	• Collect and monitor stormwater before discharge.
Indirect	
12. • Occupational health effects on workers due to fugitive dust, materials handling, noise, or other process operations.	12. Facility should implement a Safety and Health Program designed to:
	• identify, evaluate, monitor, and control health hazards
• Accidents occur at higher than normal frequency because of level of skill or labor.	• provide safety training
13. Regional solid waste problem exacerbated by inadequate on-site storage or lack of ultimate disposal facilities.	13. • Plan for adequate on-site disposal areas assuming screening for hazardous characteristics of the leachate is known.
	• Provide, in design phase, for adequate ultimate disposal facilities.

Table 10.11. Chemical and Petrochemical (continued)

Potential Negative Impacts	Mitigating Measures
Indirect (continued)	
14. Transit patterns disrupted, noise and congestion created, and pedestrian hazards aggravated by heavy trucks transporting raw materials to/from facility.	14. • Site selection can mitigate some of these problems. • Special transportation sector studies should be prepared during project feasibility to select best routes to reduce impacts. • Transporter regulation and development of emergency contingency plans to minimize risk of accidents.

FERTILIZER

1. Nearly all fertilizer manufacturing projects involve the production of compounds that supply the plant nutrients nitrogen, phosphorus and potassium, either individually ("straight" fertilizers) or in combination ("mixed" fertilizers).

2. Ammonia forms the basis for producing nitrogenous fertilizers and most manufacturing plants contain ammonia production facilities regardless of the nature of the final product. Many plants also produce nitric acid on-site as well. The preferred feedstock to produce ammonia is natural gas; however, coal, naphtha, and fuel oil are also used. The most common nitrogenous fertilizers are anhydrous ammonia, urea (produced from ammonia and carbon dioxide), ammonium nitrate (produced from ammonia and nitric acid), ammonium sulfate (produced from ammonia and sulfuric acid) and calcium ammonium nitrate or limestone ammonium nitrate (produced by adding limestone to ammonium nitrate).

3. Phosphate fertilizers include ground phosphate rock, basic slag (a by-product of iron and steel manufacturing), superphosphate (produced by reacting ground phosphate rock with sulfuric acid), triple superphosphate (produced by reacting ground phosphate rock with phosphoric acid), and mono- and diammonium phosphate (MAP and DAP). Basic raw materials are phosphate rock, sulfuric acid (usually produced on-site from elemental sulfur), and water.

4. All potassium fertilizers are produced from brines or from underground deposits of potash. The main formulations are potassium chloride, potassium sulfate, and potassium nitrate.

5. Mixed fertilizers may be produced by dry mixing, by granulation of several fertilizer intermediates mixed in solution, or by reacting phosphate rock with nitric acid (nitrophosphates).

Potential Environmental Impacts

6. This industry's positive socioeconomic impacts are obvious: fertilizer is critical to achieving the level of agricultural production needed to feed the rapidly growing world population. There are also indirect positive impacts of proper fertilizer use on the natural environment; for example, chemical fertilizer enables production to be intensified on existing cropland, while reducing the need for expansion onto lands that may have other natural or social resource values.

7. However, negative environmental impacts from fertilizer production can be severe. Wastewaters are the major problem. They may be highly acidic or alkaline and, depending on the type of plant, may contain a number of substances toxic to aquatic organisms in higher concentrations: ammonia or ammonium compounds, urea from nitrogen plants, cadmium, arsenic, and fluorine from phosphate operations where they are found as impurities in the phosphate rock. Total suspended solids, nitrate and organic nitrogen, phosphorus, potassium, and (as a consequence), elevated biochemical oxygen demand (BOD_5) are also common in effluents; and, with the exception of BOD_5, in stormwater runoff from raw material and waste storage areas. Phosphate plants can be designed to have no wastewater discharge except for evaporation pond overflow during conditions of high rainfall, but this is not always practical.

8. Finished fertilizer products are also potential water pollutants; excessive or improper use may contribute to eutrophication of surface waters and nitrogen contamination of groundwater. Phosphate mining can cause negative impacts on water quality as well. These should be considered in predicting the potential impacts of projects that will lead to new or expanded mining operations, whether or not the plant is sited at or near the mine (see the section on "Mining and Mineral Processing").

9. Air pollutants include particulate matter from boiler flues and phosphate rock grinding, fluorine (the principal air pollutant from phosphate plants), acid mist, ammonia, and oxides of sulfur and nitrogen. Solid wastes are principally from phosphate plants and consist primarily of ash (if coal is used to produce steam for the processes), and gypsum (which may be considered hazardous because of the presence of cadmium, uranium, radon gas emissions or other toxic elements in the phosphate rock).

10. Sulfuric and nitric acid manufacturing and handling constitute significant occupational safety and health hazards. Accidents resulting in ammonia releases can endanger not only plant workers, but also people who reside or work nearby. Other possible accidents are explosions and injuries to the eyes, nose, throat, and lungs.

11. Because a number of the impacts described can be avoided altogether or mitigated more successfully and at less cost by prudent plant site selection (see Table 10.12 at the end of this section), "Plant Siting and Industrial Estate Development" should be read in conjunction with this section.

Special Issues

Solid Wastes

12. Solid wastes produced during fertilizer manufacturing are complex and cannot be indiscriminately disposed of on land. Potentially hazardous materials include vanadium catalysts from sulfuric acid plants, and arsenic sludges from any sulfuric acid plants that use pyrites, and require special handling and disposal. Gypsum may be difficult to dispose of if contaminated with toxic metals. Ash from ammonia plants based on coal gasification may also be a disposal problem. Sufficient land area should be available to allow proper placement of solid wastes. Opportunities for reuse of these solid wastes exist and should be evaluated for each project (see next paragraph). Ultimate solid waste disposal measures should be identified in the project plan and thoroughly evaluated during project feasibility studies.

Waste Minimization

13. Significant quantities of water are used in the fertilizer industry for processes, cooling, and operating pollution abatement equipment. Liquid wastes originate in the processes, cooling tower, and boiler blowdown, resulting in spills, leaks, and runoff. However, there are opportunities to reuse water within plants, thereby diminishing the quantities that have to be impounded or treated and reducing the plant's demand on local water sources. For example, wastewater from phosphoric acid production can be reused as process water in the same plant. Other wastewaters can be used in condensers, gas scrubbers, and cooling systems.

14. Gypsum from phosphate fertilizer plants can be reused in cement manufacturing and production of building block and gypsum board. Gypsum has also been used as cover material in sanitary landfills. If the gypsum is contaminated with toxic metals or radioactive material, it will require special handling.

15. Hydrofluosilicic acid is widely used in the United States by water utilities that practice fluoridation because, as a waste product of phosphate fertilizer production, it is substantially less costly than sodium fluoride. Although the acid is hauled long distances on land in the United States, export is generally not economically attractive. Nevertheless, there may be circumstances in which it could be reused in a developing country, especially after converting it to sodium salt. The acid can also be used for the production of aluminum fluoride.

Ammonia

16. The production, use, and storage of ammonia will require sound design, good maintenance and monitoring to minimize the risk of accidental releases and explosions. A contingency plan to protect plant staff and neighboring communities is imperative.

Project Alternatives

Site Selection

17. General issues to consider in industrial plant siting are discussed in the "Plant Siting and Industrial Estate Development" section. The nature of fertilizer production is such that impacts on water quality and impacts of raw material extraction, and transportation of bulk materials to and from the plant warrant special attention in evaluating alternative sites. Receiving waters with substandard quality or insufficient flow to accept even well-treated effluents are inappropriate. If the demand for raw materials for a phosphate plant will necessitate opening additional quarry sites, they should be identified (if known) and their environmental impacts considered as part of the project.

Manufacturing Process

18. Although a variety of alternatives exist in project planning and execution, the type of fertilizer manufacturing process is generally constrained by the raw materials available and the demand for particular finished products. In the selection of a phosphoric acid process, the quality of the gypsum by-product could be a parameter: a hemidihydrate process could produce a gypsum that is directly usable as an additive in cement manufacture.

19. Iron and steel coking plants are an alternative but limited source of ammonium sulfate fertilizer (produced from ammonia and sulfuric acid); ammonium sulfate is a by-product of coke production and also of caprolactam (nylon) production. Natural gas, oil, naphtha, and coal are alternative raw materials for ammonia production. Sulfur and pyrites are alternatives for sulfuric acid production.

20. Natural gas, oil, and coal are alternative fuels for generating steam at fertilizer plants.

Air Pollution Controls

21.   The following measures should be considered to control emissions to the atmosphere from plant operations: design of process and choice of process equipment, electrostatic precipitators, flue gas scrubbers, baghouse filters, and cyclones.

Water Quality Controls

22.   Water pollution resulting from liquid effluent discharge or runoffs from waste piles can be controlled if properly monitored. The following options for treatment of wastewater and rinse should be included in the project design:

- wastewater reuse
- ion exchange or membrane filtration (phosphoric acid plants)
- neutralization of acidic or alkaline wastewaters
- sedimentation, flocculation, and filtration for suspended solids
- land application of process wastewaters
- biological treatment (nitrification/denitrification)

Management and Training

23.   The potential impacts of fertilizer manufacturing processes on air, water, and soil necessitate institutional support for efficient conduct and supervision of materials handling, pollution control, and waste reduction. Facility personnel should be trained in air and water pollution control technologies being employed. Equipment manufacturers will frequently supply the necessary training in equipment operations and maintenance. Standard operating procedures should be established for the plant and enforced by its management. They should include pollution control equipment operation, air and water quality monitoring requirements, instructions for operators to prevent malodorous emissions, and directives for notification of proper authorities in the event of accidental release of pollutants. Toxic and hazardous substance handling and management should be improved by detectors, alarms, etc., and special training of operating personnel.

24.   Emergency procedures are necessary to provide for rapid and effective action in the event of accidents that pose serious threats to the environment or to the surrounding community (e.g., major spills, fires, or explosions). Local government officials and agencies, as well as local community services (medical, firefighting, etc.), usually play key roles in emergencies of this sort and should be included in the planning process. Periodic drills are important components of response plans. (For further discussion, see the "Industrial Hazard Management" section.)

25.   Plant health and safety rules should be established and enforced. They should include:

- Provisions to prevent and respond to accidental ammonia releases and sulfuric, phosphoric and nitric acid spills.

- Procedures to minimize calcium ammonium nitrate explosion hazards.

- Procedures to keep exposure to ammonia, nitrogen oxide vapors (nitrogenous fertilizer plants) sulfur di- and trioxide vapors, and sulfuric acid mist below World Bank limits.

- Procedures to check for phosphoric acid filters for radioactive scalings.

- A program of routine medical examinations.

- Ongoing training in plant health and safety and in good environmental housekeeping practices.

(The following sections should be consulted in conjunction with the World Bank's <u>Occupational Health and Safety Guidelines</u>: "Industrial Hazard Management", "Hazardous Materials Management", and "Plant Siting and Industrial Estate Development.")

26. Emission and effluent standards should be set for the plant based on national regulations where they exist, or on World Bank guidelines where they do not. Government agencies charged with monitoring air and water quality, pollution control equipment operations, enforcing standards, and overseeing waste disposal activities may require specialized training and should be provided with the necessary equipment and authority. The environmental assessment should include an evaluation of local capabilities in these areas and recommend appropriate elements of assistance to be included in the project.

Monitoring

27. Monitoring plans for fertilizer plants are necessary for plant/site-specific and should include:

- continuous flue gas opacity
- periodic testing (phosphate plants only) for emissions of particulates, fluorine compounds, nitrogen oxides, sulfur dioxide
- control of sulfur oxides from sulfuric acid plants and nitrogen oxides from nitric acid plants
- periodic testing (nitrogen plants only) for particulate emissions, ammonia, and nitrogen oxides
- process parameters (continuous) that prove the operation of air pollution control equipment (e.g., flue gas temperature records will show when scrubbers were out of operation)
- workspace air quality for the following, as applicable to plant type and process: nitrogen oxides, ammonia, sulfur dioxide, fluorine compounds, and particulates
- ambient air quality in vicinity of plants for applicable pollutants
- receiving water quality downstream for dissolved oxygen and applicable pollutants
- liquid waste streams for pH (continuous), total suspended solids, total dissolved solids, ammonia, nitrate, organic nitrogen, phosphorus, BOD_5, oil and grease (when fuel oil is used)
- stormwater discharges for phosphorus, fluorine compounds, total suspended solids, and pH
- gypsum for cadmium and other heavy metals and radioactivity
- working areas of all plants for ambient noise levels
- receiving waters for pH, total suspended solids, and ambient air quality for particulate matters
- gypsum storage piles and ponds piles for runoff and infiltration
- inspection for adherence to safety and pollution control procedures, and proper maintenance programs

Table 10.12. Fertilizer

Potential Negative Impacts	Mitigating Measures
Direct: Site Selection	
1. Siting of plant on/near sensitive habitats such as mangroves, estuaries, wetlands, coral reefs.	1. • Locate plant in industrially zoned area, if possible, to minimize or concentrate the stress on local environmental services and to facilitate the monitoring of discharges. • Integrate site selection process with natural resource agencies to review alternatives.
2. Siting along water courses causing their eventual degradation.	2. • Site selection process should examine alternatives that minimize environmental effects and not preclude beneficial use of the water body. • Plants with liquid discharges should only be located on a watercourse having adequate waste-absorbing capacity.
3. Siting can cause serious air pollution problems for local area.	3. • Locate at a high elevation in an area not subject to air inversions, and where prevailing winds are towards relatively unpopulated areas.

4. Siting can aggravate solid waste problems in an area.	4. • Site selection should evaluate the location according to the following guidelines:
	  • plot size sufficient to landfill or dispose on-site
	  • proximity to suitable disposal site
	  • convenient for public/private contractors to collect and haul solid wastes for final disposal
	  • availability of options for gypsum disposal or reuse

Table 10.12. Fertilizer (continued)

Potential Negative Impacts	Mitigating Measures
Direct: Plant Operation	
5. • Water pollution from discharge of liquid effluents and process cooling water or runoff from waste piles. • Phosphate plants: phosphate, fluoride, BOD_5, Total Dissolved Solids (TDS), pH • Nitrogen plants: ammonia, urea, ammonium nitrate, COD, pH • Materials storage piles runoff: TSS, pH, metals	5. • Laboratory analysis of liquid effluent should include fluoride, BOD_5, TSS, and in-situ pH temperature monitoring. **All Plants** • No cooling water discharge. If recycling not feasible, discharge cooling water provided receiving water temperature does not rise > 3°C. • Maintain pH level of effluent discharge between 6.0 and 9.0. • Control effluent to EPA limitations (40 CFR 418) for specific process. **Material Storage Piles/Solid Waste Disposal Areas** • Minimize rainfall allowed to percolate through piles and runoff in uncontrolled fashion. • Line storage areas.
6. • Particulate emissions to the atmosphere from all plant operations.	6. • Control particulates by fabric filter collectors or electrostatic precipitators.
7. • Gaseous emission of SO_x and NO_x, ammonia, acid mist and fluorine compounds to the atmosphere.	7. • Control by scrubbing. • Analyze raw materials during feasibility stage of project. • Proper design of sulfuric acid plants and nitric acid plants with NO_x abatement equipment.

Table 10.12. Fertilizer (continued)

Potential Negative Impacts	Mitigating Measures
Direct: Plant Operation (continued)	
8. Accidental release of potentially hazardous solvents, acidic, and alkaline materials.	8. • Maintain storage and disposal areas to prevent accidental release. • Provide spill mitigation equipment. • Provide dikes around storage tanks.
9. Surface runoff of constituents, raw materials, and solid wastes frequently stored in piles on the facility grounds can pollute surface waters or percolate to ground waters.	9. • Plan proper storage in design phase. • Cover and/or line storage areas (especially gypsum piles) to prevent percolation and runoff to ground and surface waters. • Diked areas should be of sufficient size to contain an average 24 hour rainfall.
10. Occupational health effects on workers due to fugitive dust, materials handling or other process operations, and accidents occur at higher than normal frequency because of level of skill or labor.	10. • Facility should implement a Safety and Health Program designed to: • identify, evaluate, monitor, and control safety and health hazards at a specific level of detail • address the hazards to worker health and safety • propose procedures for employee protection • provide safety training
11. Regional solid waste problem exacerbated by inadequate on-site storage or lack of ultimate disposal facilities.	11. Plan for adequate on-site disposal, assuming screening for hazardous characteristics of the leachate is known.

Table 10.12. Fertilizer (continued)

Potential Negative Impacts	Mitigating Measures
Direct: Plant Operation (continued)	
12. Transit patterns disrupted, noise and congestion created, and pedestrian hazards aggravated by heavy trucks transporting raw materials to/from facility.	12. • Site selection can mitigate some of these problems. • Special transportation sector studies should be prepared during project feasibility to select best routes to reduce impacts. • Transporter regulation and development of emergency contingency plans to minimize risk of accidents.
13. Increasing nitrate pollution of ground water from use of nitrogen fertilizers.	13. Directions for use should be provided to minimize nitrate pollution potential.
14. Eutrophication of natural water systems.	14. Directions for use should be provided to minimize nitrate and phosphate pollution potential.

FOOD PROCESSING

1. Food processing projects involve the processing and packaging of meat and meat products, fish and shellfish, dairy products, fruits and vegetables, and grains. Food processing may include refinement, preservation, product improvement, storage and handling, packaging or canning.

2. The basic raw materials of the industry are either naturally produced or grown. The processing may involve receiving and storing raw or partially processed materials, processing the materials into finished products, and packaging and storing the finished products. The objective of food processing is to extend the shelf life of raw commodities through the use of various preservation methods.

Potential Environmental Impacts

3. The food processing industry provides food products for immediate or future human consumption and by-products for use in the livestock industry. The industry generates large volumes of wastewater and solid wastes and may also be a source of air pollutants. Wastewaters arise mainly from leaks, spills, and equipment washouts. Large volumes are also generated in the washing operations to remove soil, pesticides and skin from fruits and vegetables.

4. Screening is extensively employed in the industry to remove solids. The recovered solids have market value and are normally processed for animal feed. Although air emission is not a problem, odor problems can be significant. (See Table 10.13 at the end of this section for further discussion.)

Diary Products

5. The dairy processing industry manufactures 20 types of milk products including pasteurized milk, ice cream, butter, cheese, condensed milk, dry milk, whey, and cultured products. Although dairy plants may perform a combination of operations to produce several products, some plants may produce only one or two. Typical manufacturing processes for the dairy product industry involve:

- receiving and storing raw materials, comprising receiving areas, transfer equipment, and large refrigerated tanks for storage;

- clarification to remove suspended solids, and separation to remove cream -- processes usually accomplished by large centrifuges of special design;

- churning, homogenizing, culturing, condensing, and drying to produce butter, ice cream, cheese, buttermilk, etc.; and

- packaging and storing for subsequent shipment.

6. The major sources of wastes and wastewaters from the dairy processing industry are wash and rinse water from wash-ups, unrecovered by-products, spoiled or damaged products, and entrainment from evaporators.

7.   Under normal operations and with good housekeeping, receiving and storing raw materials are not major sources of waste. Solid wastes are minor and may be disposed of at a sanitary landfill.

8.   A significant characteristic of the waste streams of all dairy plants is the marked variations in flow, BOD_5, temperature, and pH. In a fluid milk plant, approximately 94 percent of the BOD_5 is contributed by milk, milk products and other edible products. Of all the wastes, whey presents the most difficult disposal problem. The most common disposal methods are those used in livestock feed and spray irrigation, discharge to municipal systems, concentration, and drying.

9.   The main safety hazards in the dairy industry result from bursting bottles, flying glass, and falls on slippery floors. Common health hazards include animal diseases such as brucellosis, bovine tuberculosis, anthrax, etc. Workers may also contract cheese-workers' itch.

Fruit and Vegetable Processing

10.   Canning and preserving extends the shelf life of raw commodities. Preservation methods involve canning, freezing, dehydrating, and brining. Fruit and vegetable preserving generally includes cleaning, sorting, peeling, sizing, stabilizing, and processing.

11.   Prior to processing, the fruits and vegetables are washed and rinsed using great quantities of water and occasionally detergents. The washed products are sorted and graded by mechanical, optical, manual or hydraulic means. Mature products are separated using a brine solution of controlled density. Following sorting, the products are mechanically stemmed, snipped, and trimmed.

12.   Many fruits and vegetables are peeled to remove residual soil, pesticides, and coarse, fuzzy or tough peels. The process is accomplished either mechanically, thermally or chemically. Pitting, coring, slicing, and dicing are carried out mechanically without the use of water. Some fruits are pureed and squeezed to produce juices. Vegetables, on the other hand, are blanched and canned. Finally, depending on the type of operation, some products are dried or dehydrated, some are cooked and others are freeze-dried.

13.   Fruits and vegetables processing plants are major water users and waste generators. The washing and rinsing operation, sorting, in-plant transport, peeling, blanching, canning, mixing, cooking and clean-up are major generators of wastewaters and solid wastes. Gaseous emissions are minor though odors may be significant in some cases.

14.   The wastewater parameters of significance are BOD_5, TSS and pH. Fecal coliforms may be of concern but can be prevented by practicing good housekeeping and maintaining sanitary conditions at all times. Because of the wide variation in flow and strength (BOD_5) of the wastewaters, treatment facilities must be designed to handle large, intermittent volumes. Citrus wastes contain pectin substances that interfere with settling of suspended solids.

15.   In fruit and vegetable canneries, the major accidents are due to lifting, burns from steam, acids and alkalis, and cuts from broken glasses and sharp tins. The main health problem is dermatitis and skin infection caused by chemicals, and the handling of fruits and vegetables. In some plants, excessive noise, temperature stress, and high humidity may also create health problems.

Meat Processing

16. Meat processing plants purchase animal carcasses, meat parts and other materials. They also manufacture sausages, cooked meats, cured meats, smoked meats, canned meats, frozen and fresh meat cuts, natural sausage casings, and other specialties. The processing operation may be carried out separately or in conjunction with slaughterhouses.

17. The meat processor receives frozen carcasses that are wet or dry thawed, or chipped. Unlike dry thawing, wet thawing generates large volumes of wastewater. Chipping involves size reduction equipment designed to handle frozen meat. A typical plant may involve one or more of the following operations:

- meat cutting to prepare standardized products for hotels, restaurants, institutions, fast food outlets, etc.;

- ham processing for curing in pickling solutions followed by cooking, smoking, cooling, slicing, and packaging;

- sausages and luncheon meats manufacturing require substantial size reduction, intensive mixing and molding, or forming of the finished product; and

- canned products for hams, sandwich spread, and pet foods.

18. Meat processing is a year-round operation with daily operation on an intermittent basis. Plants usually shut down daily for extensive clean-ups. The industry produces large volumes of wastewaters with varying amounts of suspended solids. Solid wastes, resulting mainly from screening and housekeeping, are normally recovered and sent to a rendering plant. Although gaseous emissions are not significant, odors are a problem. They originate from cooking of animal materials, animal residues, and decomposition of organic matter.

19. The most important parameters of concern in the meat processing industry are BOD_5, TSS, oils and grease, pH and fecal coliforms. Depending on the type of operation, phosphorous and ammonia may also be a problem. Of the processes described previously, meat canning and ham processing are the largest contributors to wastewater flows, BOD_5, TSS, and oil and grease. Meat cutting operations are the lowest.

20. The wasteload discharged from the meat industry can be reduced to desired levels through effective water management, in-plant waste controls, process modifications, and wastewater treatment systems.

21. Safety hazards in the meat processing industry result mainly from slippery floors, burns, and cuts and abrasions from sharp tins, broken glasses and cutting machinery. Main health problems are dermatitis caused by chemicals and skin infections. Diseases associated with animals, such as anthrax, actinomycosis, erysipeloid and tuberculosis, are also a potential source of health hazard. Other health problems may include noise, high temperature, and humidity.

Fish and Shellfish Processing

22. The canned and preserved fish and seafood industry has progressed steadily from drying and curing techniques to preserving, canning, freezing, and rendering of fish products. The length of the fish processing season varies greatly depending upon the harvesting season and the amount of material processed in the industry.

23. The process used in this industry includes harvesting of the fish, storing, receiving, eviscerating, pre-cooking, pick-up or cleaning, preserving and packaging. Following harvesting, fish is unloaded from the vessel, weighed and transported to the processing area for either immediate processing or cold storage. In some operations, pre-processing to behead shrimp, eviscerate fish or shell fish is carried out at sea. The wastes are dry captured or screened from wastewaters and processed as a by-product.

24. Depending on the final product destination, fresh fish and seafood may be packaged for immediate consumption or cooked for a picking and cleaning operation to remove skin, bone, shell, gills, etc. The picking may be followed by freezing, canning, pasteurization, and refrigeration.

25. There is considerable variation, from process plant to process plant, in the amount of water used and the waste generated. In general, wastes from this industry contain BOD_5, COD, TSS, oil and grease, and may be of high or low pH. Normally, these wastewaters contain no hazardous or toxic materials. Occasionally, wastewaters containing high concentrations of sodium chloride may be discharged.

26. Under normal operations, gaseous emissions are not a problem. Solid wastes, if not recovered, could present a treatment and disposal problem. Fortunately, newer plants recover most solid wastes by screening or dry collection. These wastes are further processed as fish meal, concentrated protein solubles, oils, liquid fertilizers, fish food pellets, animal feed, shell novelties, etc.

27. In the fish canning industry, major accidents result from lifting, handling, and falling materials. Secondary causes include falls on slippery floors and burns and cuts from machinery and sharp objects. The main health problems arise from warts due to virus and fish slime, and skin infections and dermatitis caused by chemicals.

Natural Resource Issues

Water

28. Significant quantities of water are used in the food processing industry. The main uses of the water are for washing and rinsing, and in-plant transport of products, and clean-ups.

29. In the fruit and vegetable industry, for example, it has been common to use water to transport the raw materials within a plant and to consider such use as both economical and of sanitary significance. However, the leaching of solubles from the products (e.g., sugars and acids from cut fruits and sugars and starch from cut vegetables) has lead to alternative means of fluid transport, such as osmotically equivalent fluid systems. Nonetheless, effective washing after harvesting is required due to use of pesticides and other contaminants and mechanical harvesting techniques that result in residues of soil and dirt in the fruits and vegetables.

30. Dairy, meat, fish, and shellfish processing also require large volumes of fresh water in the process and for clean-up of the equipment and process areas. Water also serves as a solvent for products and as a medium for cooking and cleaning. Therefore, the siting of the food processing facilities must be such that water supplies of adequate quality and quantity are available.

31. Wastewater streams for the food processing industry vary with the type and size of the processing operation. Effluents typically have high BOD_5, COD, oil and grease, coliforms, and suspended and dissolved solids. Other contaminants such as pesticide residues, complex oils, alkaline or acidic compounds and other organic constituents may also be present in the wastewater effluents. The World Bank's Environmental Guidelines and Occupational Health and Safety Guidelines provide effluent and worker safety standards for the food processing industries.

32. The U.S. Environmental Protection Agency (EPA) has also established effluent guidelines for the various sectors of the food processing industry. Compounds that are regulated under EPA standards for various food processing operations include pH, total suspended solids (TSS), and BOD. Standards for oil and grease, fecal coliform, and ammonia have also been established for the dairy, meat, and seafood processing industries. National regulations vary by country and industry type and can be highly subjective.

33. Water resources in the surrounding area can be negatively affected by accidental releases of untreated wastewaters and processing chemicals, or through inadequate control of surface runoff and other non-point sources. Where chemicals are used, chemical handling and storage procedures, and spill control measures should be developed to minimize the potential for accidental release to the environment.

Air

34. Air emissions from food processing facilities are minimal but may include particulate matter, sulfur oxides, nitrogen oxides, hydrocarbons and other organic compounds. Noxious and nuisance odors are a major problem associated with food processing industries. The U.S. Environmental Protection Agency has developed air quality standards for particulate matter, sulfur oxides, and a number of organic compounds. In the absence of local standards, USEPA standards may be appropriate for food processing facilities in other areas of the world.

Land

35. Location of food processing and manufacturing facilities may adversely affect land resources through the development of ecologically, agriculturally or economically important land. Land resources may also be adversely affected through the disposal of solid wastes on the property. Sufficient land areas must be acquired to allow for logical and unrestricted placement of processing and storage facilities.

36. Where they exist, local regulations should be followed. Measures to minimize the contamination potential from solid waste disposal should also be reviewed and incorporated into the development plan. (The section on "Solid Waste Collection and Disposal Systems" provides more information on this subject.)

Sociocultural Issues

37. When properly designed and operated, food processing and manufacturing facilities provide local employment opportunities and an outlet for local produce with minimal effect on the environment. However, facilities that are inadequately designed and operated can result in adverse impacts on the local cultural resources, cause nuisance and health problems from noxious and odorous emissions, reduce land values, as well as degrade air, land, and water resources. The damage or degradation can limit the potential for additional development. During project appraisal (feasibility study stages), an evaluation should be made of the proposed site, with respect to the above factors, access to social amenities, availability of a skilled labor force, and support and service industries. Noise and odorous emissions from the processing facilities should also be evaluated since these may effect employee health and affect the local communities.

Special Issues

Environmental Law and Land Use

38. Facility siting is a highly complex and time consuming process involving developers, special interest groups, politicians, and local and national authorities. The time and cost of obtaining approval for siting of food processing facilities should be factored into the overall project schedule and cost. Developing countries are becoming more experienced in the facility siting process and many have applicable land-use and environmental regulations; therefore, integrated planning of facility siting and operation is important.

Solid Waste and Residue Utilization

39. Much of the solid waste materials produced from food processing facilities is meat by-products, fruit and vegetable matter, and waste fish and fish by-products. Meat and fish by-products should not be disposed of in landfills because they may become potential habitat for disease carrying vectors (rats, insects, etc.). These materials present valuable resources that should be recovered. The meat by-products should be collected and delivered to a rendering plant for processing into lard, tallow, oils, proteinaceous solids, and bone meal for fertilizers or other products. In fruit processing, peel and core particles along with inferior fruit can be used to make juice or vinegar stock. Peelings of citrus fruits can be reprocessed to recover citrus oil. In potato processing, the peelings can be processed to recover crude starch. Other wastes can be used as cattle feed, composted or disposed of in a landfill.

40. Solid wastes from small seafood processing operations are commonly disposed of in waterways near the processing facility. This should be avoided for large operations and in areas not adjacent to large water bodies or where the current is insufficient to carry the waste products out to sea. In the latter case, the solids will settle and break down anaerobically, and that can lead to objectionable odor problems. The solid waste by-products of fishery operation can be recovered and processed to produce fish meal, concentrated protein solubles, oils, liquid fertilizer, fish feed pellets, shell novelties and pearl essence.

Process Specifications

41. Food processing facilities should develop specifications to minimize the potential for improper food preparation that could result in food-related illnesses (e.g., botulism). These specifications should be a combination of:

- improved management controls and measures to minimize product losses, maintain equipment, and develop alternative uses for wasted products;

- improved engineering and process equipment to increase production efficiency and reduce waste loads; and

- improved sanitary conditions to eliminate the potential for bacterial contamination by using proper cooking time, appropriate equipment, and cleaning procedures.

Project Alternatives

42. Although a variety of alternatives to project design and execution exist, the types of food processing and manufacturing facilities are limited to the available technologies, raw materials, and markets for the manufactured products. The alternatives listed below provide a framework that could be used in preparing and reviewing environmental assessment, in evaluating specific projects, and in assisting the project design team.

Site Selection

43. Selection of a site for food processing and manufacturing facilities is dependent on a number of economic, ecologic, and sociopolitical concerns. Regardless of the product processed or manufactured, an environmentally ideal site is one that satisfies the following criteria.

- sufficient land area to provide planned and unrestricted development of facilities for storage of raw materials and processing, manufacturing, and waste disposal facilities;

- minimal displacement of people or houses;

- minimal conflicts with higher valued land uses such as agriculture, especially in marginal land areas where prime agricultural lands may be at a premium;

- proximity to receiving waters for effluent discharge without significant impact to the biophysical and aquatic environment;

- easy access to social and physical infrastructures such as a skilled labor pool, support industries, a transportation network, an energy supply, raw materials, and a potential market for products;

- sufficient distance from tourist or recreational areas and office buildings and housing complexes to minimize the impacts of odors, noise and other pollutants; and

- minimal impact during construction and operation of the facilities on rare, threatened, or endangered species and their habitats.

Raw Material Supply

44. It is necessary to ensure that the raw materials used in food processing and manufacturing are supplied in a hygienic and environmentally sound manner by minimizing impacts on other sectors, habitats, and resources. Seafood products, for example, should not be harvested from areas where fish populations are stressed or where existing pollution may have contaminated the harvest. Similarly, products that may have been contaminated with pesticides or other chemicals, or products that have been improperly stored, should not be processed for human consumption.

Operation of the Facility

45. Food processing and manufacturing uses a variety of processes. The types of end product and the size of the operation determine the type of equipment used and the quality and quantity of waste produced. This in turn determines the necessary type of pollution control equipment. The equipment used in pollution abatement cannot be specified for all possible food processing and manufacturing industries. General pollution treatment measures used in the industry include:

(a) **Water Pollution**

- activated sludge treatment
- aerated lagoons
- screening
- sedimentation, flocculation, neutralization, clarification
- spray irrigation
- trickling filters
- stabilization lagoons
- air flotation
- ammonia stripping
- ion exchange
- carbon adsorption
- electrodialysis

(b) **Air Pollution**

- electrostatic precipitators baghouses
- activated charcoal filtration
- scrubbing with sodium hypochlorite (for odor control)

Management and Training

46. The need for proper management and training becomes more important as the technical sophistication of a process increases. In developing countries, it is increasingly important to develop the technical capabilities of industry personnel and for government officials to monitor pollution abatement controls. The number of in-country consultants qualified to prepare detailed environmental assessments is usually small and many of these are academics who lack project or industrial experience.

47. Government and industry alike need to develop expertise to select EA contractors and consultants who can provide competent and cost-effective services during planning, design, construction/operation, and maintenance of the project and associated facilities. Sometimes retired consultants who have established reputations in one or more assessment areas may serve as independent advisors to host governments or industry, and assist in optimizing conservation and environmental training components of development projects.

48. To improve the quality of projects by employing environmental management objectives, several areas may require strengthening.

- Training for government professional staff EAs, analyzing and interpreting pollution data, and taking enforcement actions where appropriate.

- Training for industry employees to develop awareness and the capability to evaluate government regulations, pollution data, treatment options and operational data.

- Training for local professionals, not associated with government and industry, to provide consulting services and/or independent review for EAs and pollution abatement.

49. Specification in the terms of reference for projects should allow consultants flexibility to design the training component according to the aptitudes of indigenous trainees and the level of experience in a country or region.

50. Institutional support is necessary for efficient operation of pollution control and waste reduction strategies. Facility personnel should be trained in air and water pollution control technologies as well as in the operation of equipment used at the facility. Pollution control equipment manufacturers and suppliers will frequently provide training in equipment operations and maintenance. In-house training, in-plant health and safety procedures, and good environmental housekeeping practices are also recommended.

51. Employees should receive training under "standards of practice" for occupational health and safety, if local regulations do not exist. Health training should emphasize the need for sanitary and sterile conditions during the handling and processing of food products to minimize disease transmission. Training should also provide information on hazards associated with chemical uses and the operation of processing equipments.

52. In addition to employee training, it may be necessary to develop training programs for personnel from local or national environmental agencies. Such training programs should cover potential environmental and health hazards associated with food processing industries and measures to minimize them. Regulatory personnel should also be trained in the collection and evaluation of environmental, health, and safety monitoring data.

Monitoring

Facility Monitoring during Start-up and Operation

53. Monitoring of air, liquid effluent, and solid wastes is generally required for pollution control in food processing projects. The development and implementation of an Environmental Monitoring Plan provides specific means to determine whether or not the project or any of its subcomponents complies with the applicable environmental standards and practices.

54. At a minimum, the monitoring plan should lay out the institutional and administrative means and schedule for surveillance and supervision of the environmental components (e.g., pollution abatement) of a project. In addition to the monitoring plan, monitoring data may be used to gain assistance from local and foreign environmental professionals at critical milestones of the project. One option is to stage workshops to evaluate environmental monitoring data, fine tune project objectives, and develop more practical management guidelines.

55. The following are representative activities to be included in the monitoring plan for food processing facilities.

- Monitoring of waste streams and gaseous emissions for selected parameters.

- Where a particular discharge has consistently high values above the national emissions limit or the standard for the industry, corrective action should be taken.

- Corrective action may involve process or equipment modification, upgrading, or housekeeping changes.

- Monitoring of receiving water quality and air quality downwind of plant.

- Monitoring of the effects of solid waste practices on land, groundwater, and surface water resources.

- Incorporation of programs to enhance environmental awareness of all employees.

- Periodic review of technology so as to adopt the most efficient and cost-effective pollution abatement systems where feasible.

- Motivation of plant managers and facility engineers to be vigilant about the potential effects of the facility on the local environment.

- Develope and maintain an odor complaints and response system to be reviewed with officials and the communities.

- Implementation of health and safety plans and regular site inspections to ensure that the training protocols and worker protective equipment are being employed in the workplace.

- Standard industry practices should be followed.

- Documentation and records should reflect periodic review and corrective actions taken.

56. An important factor for pollution abatement in food processing projects is the simultaneous strengthening of both in-plant and government monitoring capabilities of legal and regulatory enforcement. Likewise, the technical capability to comply with effluent standards is necessary. To operate a successful monitoring program, it may be necessary to introduce sampling equipment and laboratory protocols (or the analytical laboratory) to the host country and incorporate the necessary training into the project design.

SMALL- AND MEDIUM-SCALE INDUSTRIES

The first edition of the Environmental Assessment Sourcebook will not discuss small- and medium-scale industrial projects. However, this section will be expanded in future revisions as information and experience are acquired. Most of the environmental issues which pertain to this sector are addressed in the sections on industrial projects. The next edition will focus on the institutional and strategic dimensions of managing the environmental impacts from small- and medium-scale industries.

Table 10.13. Food Processing

Potential Negative Impacts	Mitigating Measures
Direct: Site Selection	
1. Siting of plant on/near sensitive habitats such as mangroves, estuaries, wetlands, coral reefs or use of prime agricultural lands.	1. • Locate plant to minimize or concentrate the stress on local environmental services and to facilitate the monitoring of discharges. • Integrate site selection process with natural resource agencies to review alternatives.
2. Siting along water courses causing their eventual degradation.	2. • Site selection process should examine alternatives that minimize environmental effects and not preclude beneficial use of the water body using the following guidelines: • on a watercourse having maximum water dilution and absorbing capacity • in an area where wastewater can be reused with minimal treatment for agricultural or industrial purposes • within a municipality that is able to accept the plant wastes in their sewage treatment system
3. Siting can cause serious odor pollution problems for local area.	3. • Locate plant in an area not subject to air inversions or to trapping pollution, and where prevailing winds are towards relatively unpopulated areas.
4. Siting can aggravate solid waste problems in an area.	4. • For facilities producing large volume of waste, site selection should evaluate the location according to the following guidelines: • plot size sufficient to landfill or on-site disposal • proximity to a suitable disposal site • convenient for public/private contractors to collect and haul solid wastes for final disposal

Table 10.13. Food Processing (continued)

Potential Negative Impacts	Mitigating Measures
Direct: Plant Operation	
5. • Water pollution from discharge of liquid effluents and process cooling water or runoff from waste piles. • Plant: Oil and Grease, TDS, TSS, BOD, COD	5. • Laboratory analysis of liquid effluent should include oil and grease, TDS, TSS, BOD, COD and in-situ temperature monitoring. **All Plants** • No cooling water discharge. If recycling not feasible, discharge cooling water provided receiving water temperature does not rise > 3°C. • Maintain pH level of effluent discharge between 6.0 and 9.0. • Control effluent to EPA specified limitations (40 CFR 405-409; 432) for specific process. • Land application of waste effluents where appropriate.
6. Particulate emissions to the atmosphere from all plant operations.	6. Control particulates by fabric filter collectors or electrostatic precipitators.
7. Gaseous and odorous emissions to the atmosphere from processing operations.	7. • Control by natural scrubbing action of alkaline materials. • Analysis of raw materials during feasibility stage of project can determine levels of sulfur to properly design emission control equipment.
8. Accidental release of potentially hazardous solvents, acidic alkaline materials.	8. • Maintain storage and disposal areas to prevent accidental release. • Provide spill control equipment.

Table 10.13. Food Processing (continued)

Potential Negative Impacts	Mitigating Measures

Indirect

9. - Occupational health effects on workers due to materials handling, noise or other process operations.
   - Accidents occur at higher than normal frequency because of level of knowledge and skill.

9. - Facility should implement a Safety and Health Program designed to identify, evaluate, monitor, and control safety and health hazards at a specific level of detail, and to address the hazards to worker health and safety and procedures for employee protection, including any or all of the following:
   - site characterization and analysis
   - site control
   - training
   - medical surveillance
   - engineering controls, work practices and personal protective equipment
   - monitoring
   - informational programs
   - handling raw and process materials
   - decontamination procedures
   - emergency response
   - illumination
   - sanitation at permanent and temporary facilities
   - regular safety meetings

10. Regional solid waste problem exacerbated by inadequate on-site storage.

10. Plan for adequate on-site disposal areas assuming that the characteristics of the leachate is known.

Table 10.13. Food Processing (continued)

Potential Negative Impacts	Mitigating Measures
Indirect (continued)	
11. Transit patterns disrupted, noise and congestion created, and pedestrian hazards aggravated by heavy trucks transporting raw materials to/from facility.	11. • Site selection can mitigate some of these problems, such as pedestrian hazards. • Special transportation sector studies should be prepared during project feasibility to select best routes to reduce impacts. • Transport regulation and development of emergency contingency plans to minimize risk of accidents.

12. Potential for disease transmission from inadequate waste disposal.	12. • Develop specifications for:
	  • food preparation and or processing
	  • waste disposal processes
	  • monitor fecal coliform or other bacteria

IRON AND STEEL MANUFACTURING

1. Iron and steel manufacturing involves a series of complex processes whereby iron ore is converted into steel products using coke and limestone. The conversion processes comprise the following steps: (a) coke production from coal and by-product recovery, (b) ore preparation (e.g., sintering, pelletizing), (c) iron production, (d) steel production and (e) casting, rolling, and finishing. These steps may be integrated at one facility or completed separately at various locations. In many developing countries, steel is manufactured from scrap products in electric arc furnaces. Steps (a) through (c) may therefore not always be applicable to all steel manufacturing projects. An alternative way to produce steel is by direct reduction with natural gas and hydrogen. The resulting product, sponge iron, is converted to steel in an electric arc furnace, followed by billet casting and one or two rolling mills for the production of non-flat products. These are the so-called mini-mills.

Potential Environmental Impacts

2. The steel industry is one of the most basic industries in developed and developing countries. In the latter this industry often provides a cornerstone for the whole industrial sector. Its economic impact is of great importance as an employer and, as a supplier of basic products, to a multitude of other industries: building and construction, machinery and equipment, and the manufacturing of transport vehicles, and railways.

3. Considerable quantities of wastewater and air emissions are generated in the course of making iron and steel. If not adequately managed, significant degradation of land, water, and air can result (see Table 10.14 at the end of this section for further discussion). A brief description of the waste generated during the iron and steel making processes is given in the following sections.

Coke Production and By-Product Recovery

4. Coke is produced by heating bituminous coal to drive off volatile components. Coke is used in iron producing blast furnaces as a reducing agent to convert iron ore to iron, and from this process a certain amount of carbon from the coke is dissolved in the liquid iron. During the coking process, great amounts of gas containing carbon monoxide are evolved, thereby furthering a whole series of chemicals: coal tar, crude light oils (containing benzene, toluene, xylene), ammonia, naphthalene, and significant amounts of water vapor. Most of these substances can be recovered and refined as chemical products; the remaining coke oven gas is used internally in different processes and furnaces for heating purposes, while surplus gas could be used either for power generation or as raw material for the production of chemicals.

5. Coke production generates significant quantities of wastewater which contains ammonia and other components released in the coking process. This water contains potentially toxic concentrations of phenols, cyanide, thiocyanate, ammonia, sulfide, and chloride. Air emissions from coke production include visible smoke, coke dust, and most of the volatile substances mentioned above.

Ore Preparation

6. Iron bearing ores (hematite, limonite, magnetite) are crushed, sized, and agglomerated through sintering, pelletizing, noduling, and briquetting to produce a preconditioned concentrated ore feeding the blast furnace process. Ore preparation can generate large quantities of tailings and results in emission of dust and sulfur dioxide.

Iron Production

7. Iron production occurs in a blast furnace and involves the conversion of iron ores into molten iron by reduction with coke and separating undesirable components such as phosphorus, sulfur, and manganese through the addition of limestone. Blast furnace gases are significant sources of particulate emissions and contain carbon monoxide. Blast furnace slag is formed by the reaction of limestone with other components and silicates present in the ores. The slag is quenched in water that may result in emissions of carbon monoxide and hydrogen sulfide. Liquid wastes from iron production originate from the scrubbing of flue gases and slag quenching. This wastewater is commonly high in suspended solids and may contain a wide range of organic compounds (phenols and cresols), ammonia, arsenic compounds and sulfides.

Steel Production

8. The iron produced in the blast furnaces is refined in the steel process where most of the carbon dissolved in the molten iron is removed. In old plants, the steelmaking process still uses the open hearth furnace, but in new plants the most favored method is the basic oxygen furnace in which carbon is dissolved in the liquid iron and burned off with oxygen. In both processes, considerable amounts of hot off-gases contain carbon monoxide and dust. These gases can be recycled after dedusting.

Casting, Rolling, and Finishing

9. The final steps in steel production involve the shaping of steel ingots or billets into desired end-products. Ingots or billets are rolled into slabs, wire, sheets, plates, bars, pipes, and rods. During rolling, large quantities of lubricating and hydraulic oils are used. In addition, pickling (cleaning to remove oxides) and cleaning of the final product to remove oil and grease may generate significant quantities of acidic, alkaline, and solvent liquid wastes. In modern plants, the step of ingot casting is often omitted and the molten steel is used directly in continuous casting and milling.

Direct Reduction: Steel Mini-mills

10. The integrated mini-mill consists of a direct reduction furnace and an electric arc furnace with continuous billet casting. It is here that reduction of the iron ore is achieved by the use of natural gas (or oil products) which is converted in a gas reforming furnace into a hydrogen containing gas. The sponge iron produced in the reduction process is fed to an electric arc furnace for conversion to steel. In this furnace, in addition to the sponge iron, considerable quantities of scrap iron are often used. The omission of the coking process and the use of high grade iron ores makes this process alternative less polluting than the conventional blast furnace process, but there can still be a significant emission of dust and carbon monoxide.

Special Issues

Solid Waste

11. Iron and steel works produce considerable amounts of solid waste, such as blast furnace slag, which can be used to produce certain cement qualities if properly granulated. Basic slag, another solid waste, can be used as fertilizer and is formed when high phosphorus iron ores are used.

12. Dust collection in the coke and sinter plants and the blast furnace all produce waste products that can be partly recycled, in principle. The design should maximize the recycling of solid waste collected from thickeners, settling tanks, dust cyclones, electrostatic precipitators, and from storage areas of raw materials. Ultimate solid waste disposal measures should be identified in the project plan and thoroughly evaluated during project feasibility studies. These waste products should be tested on leachability, and solid disposal areas should be lined and monitored continuously on groundwater pollution. (See the section on "Industrial Hazard Management".)

Liquid Waste

13. Solvents and acids used in cleaning steel are potentially hazardous substances and should be handled, stored, and disposed of as hazardous substances. Several of the by-products to be recovered are either hazardous or cancer causing agents and adequate measures should be taken to collect, store, and dispatch these products. The monitoring of liquid and gaseous leakage is required.

Waste Minimization

14. If appropriate measures are not taken, air pollution can be a very serious problem. It will be necessary in the design stage to study minimization of air pollution through special equipment for dry dust removal, for scrubbing of the off-gases to recover valuable chemicals, and to remove toxic contaminants and for collecting gases containing carbon monoxide and hydrogen to be used as secondary fuel in the plant or as a base for producing other chemicals (e.g., methanol and ammonia). Such measures can reduce air pollution and increase energy efficiency. Chemicals causing air pollution are sulfur dioxide, nitrogen oxides, benzene, toluene, xylene, naphthalene, phenols, benzopyrene, cyanide, hydrogen sulfide, and lead and zinc compounds.

15. Significant quantities of water are used in iron and steel manufacturing. Wastewater treatment systems are required for all iron and steel making processes and recycling of used and treated water should be considered. Because of the high solids content of used water for scrubbing, extensive coagulation and settling facilities are required.

16. The World Bank's Environmental Guidelines provides emission standards; air quality standards and water discharge are regulated by the U.S. Environmental Protection Agency (EPA). These regulations can be used as guidelines for iron and steel manufacturing projects in developing countries. Appropriate storage practices for liquids may include the use of double walled tanks or diking; in addition, leak detection systems are required for both liquids and gases as well as for tanks and piping. (Further information on this subject is provided in the section on "Industrial Hazard Management.")

Project Alternatives

Site Selection

17. General issues for consideration in industrial plant siting are discussed in the section on "Plant Siting and Industrial Estate Development." The nature of the iron and steel production is such that impacts on the environment from production, storage, and transportation warrant special attention in evaluating alternative sites. The impact on the environment can be substantial if insufficient attention has been given to emission and effluent problems in the planning stage. Receiving waters with substandard quality or insufficient flow to accept even well-treated effluents are inappropriate.

18. Another aspect that needs attention is the transport of raw materials to the site and the end product away from the site. The siting of industrial plants near living areas, especially when densely populated, should be avoided because of dust and noise nuisance. Iron and steel work requires a spacious lay-out and the site selection should be governed accordingly. Also, an attempt should be made to provide space for additional facilities in the future.

Manufacturing Process

19. Although a variety of alternatives exist in project planning and implementation, the iron and steel process is generally constrained by the raw materials available, such as iron ore, that can vary greatly in mineral, chemical, and physical properties by the raw materials used for reduction in the blast furnace (like cokes with additional natural gas, oil or coal fines injection), and by the fuels used for furnaces, boilers and power stations. The choice of final products also influences the design of the plant. A steel mini-mill with iron ore direct reduction and an electro-furnace based on natural gas and electricity will have considerably less environmental impact. The recent designs of integrated iron and steel plants show the trend toward a continuous process of less cooling and heating at interfaces -- important for energy savings -- and decreased air and water pollution.

20. There is a wide range of pollution control processes and equipment available to choose from. The choice method of control and equipment will vary the volume and composition of pollutants to be recovered or discharged to the environment.

Air Pollution Controls

- electrostatic precipitators
- types of cyclones
- adequate pelletizing of fines
- gas coolers, venturi scrubbers, and separators
- flue gas scrubbing
- ammonia, benzene, and hydrogen sulfide recovery equipment
- sulfur dioxide recovery equipment
- bag filters
- carbon monoxide recovery and recycling
- waste heat recovery

Water Quality Controls

- neutralization of acidic and alkaline waste streams
- sedimentation, flocculation in thickeners
- filtration for remaining suspended solids
- oil and water separators
- control of organic content with active carbon treatment
- ion exchange for control of metals
- reverse osmosis for control of metals
- water reuse and recycling or evaporation with waste heat

Management and Training

21. Institutional support for efficient management of pollution control and waste reduction strategies may be required for iron and steel projects to minimize the potential negative impact of such a manufacturing complex on air and water quality. Plant staff should include a plant engineer trained in the monitoring of air and water pollution control technologies. If requested, manufacturers will frequently provide training in equipment operations and maintenance. Standard operating and predictive maintenance procedures should be established for the plant and enforced by management. They should include pollution control equipment operation, air and water monitoring requirements, and instructions for notification and shutdown or other responses to pollution equipment failure.

22. Plant health and safety rules should be established and enforced. In addition to the normal rules, they should include:

- Provisions to prevent and respond to hazardous gases (such as carbon monoxide and ammonia) in enclosed areas, and spills of hazardous fluids (such as sulfuric acid).

- Provisions to minimize exposure to noise and excessive heat hazards associated with the operation of steel production involving heavy equipment.

- A program of routine medical examinations.

- Ongoing training in plant health and safety, and in environmental housekeeping practices.

- Emergency procedures requring regular drills to provide for a plan of action in case of a major leak, spill, explosion or fire.

(See also the "Industrial Hazard Management" section and the World Bank's <u>Occupational Health and Safety Guidelines</u>.)

23. Emission and effluent standards should be set for the plant based on national regulations where they exist or on World Bank guidelines where they do not. Government agencies charged with monitoring the operation of pollution control equipment, air and water quality, enforcing standards, and overseeing

waste disposal activities should have the necessary equipment and authority to do so. Specialized training may also be required. The environmental assessment should include an evaluation of local capabilities in these areas and recommend appropriate elements of assistance to be included in the project.

Monitoring

24. Monitoring plans are necessary for plant- and site-specific. In general, plans for iron and steel works should include monitoring for:

- emissions of particulates, sulfur dioxide, carbon monoxide, ammonia, hydrogen sulfide, arsenic and cyanides
- process parameters proving adequate operation of air pollution abatement equipment
- flue gas opacities and combustion efficiency (boiler house, power generation)
- workspace air quality as applicable to plant type and process on particulates, sulfur dioxide, and nitrogen oxides
- ambient air quality downwind in the vicinity of plants for pollutants, and particulates
- receiving water quality downstream for dissolved oxygen, pH, applicable pollutants, and suspended solids
- wastewater streams from plants and sedimentation tanks for suspended solids, pH, applicable pollutants, BOD5, oil and grease
- stormwater discharge on oil and grease and suspended solids
- effects of solid waste storage practices on ground and surface water
- working areas of all plants for ambient noise levels
- outside plant noise levels
- adherence to safety and pollution control measures

Table 10.14. Iron and Steel Manufacturing

Potential Negative Impacts	Mitigating Measures
Direct: Site Selection	
1. Siting of plant on/near sensitive habitats such as mangroves, estuaries, wetlands, coral reefs.	1. • Locate plant in industrially zoned area, if possible, to minimize or concentrate the stress on local environmental services and to facilitate the monitoring of discharges. • Involve natural resource agencies in site selection process to review alternatives.
2. Siting along water courses causing their eventual degradation.	2. • Site selection process should examine alternatives that minimize environmental effects and do not preclude beneficial use of water bodies. • Plants with liquid discharges should only be located on a water-course having adequate capacity to assimilate waste in treated effluent.
3. Siting can cause serious air pollution problems for local area.	3. Locate plant at elevation above local topography, in an area not subject to air inversions, and where prevailing winds are towards relatively unpopulated areas.
4. Siting can aggravate solid waste problems in an area.	4. Site selection should evaluate the location according to the following guidelines: • proximity to suitable disposal site • plot size sufficient for landfill or disposal on-site • convenient for public/private contractors to collect and haul solid wastes for final disposal • reuse or recycle materials to reduce waste volumes

Table 10.14. Iron and Steel Manufacturing (continued)

Potential Negative Impacts	Mitigating Measures
Direct: Plant Operation	
5. • Water pollution from discharge of liquid effluents and process cooling water or runoff from waste piles. • Plant: Total Suspended Solids (TSS), oil and grease, ammonia nitrogen, cyanide, phenols, benzene, naphthalene, benzo-a-pyrene, pH, lead, zinc • Materials storage piles runoff: TSS, pH, metals	5. • Laboratory analysis of liquid effluent should include: TSS, oil and grease, ammonia nitrogen, cyanide, phenols, benzene, naphthalene, benzo-a-pyrene, pH, lead, zinc, and in-situ temperature monitoring. _All Plants_ • No cooling water discharge. If recycling not feasible, discharge cooling water provided receiving water temperature does not rise >3°C. • Maintain pH level of effluent discharge between 6.0 and 9.0. • Control effluent to specified limitations in Bank or other guidelines (e.g., EPA 40 CFR 420) for specific process. _Material Storage Piles/Solid Waste Disposal Areas_ • Minimize stormwater allowed to percolate through materials and runoff in uncontrolled fashion. • Line open storage areas.
6. Particulate emissions to the atmosphere from all plant operations.	6. Control particulates by fabric filter collectors or electrostatic precipitators.
7. Gaseous emission of SO_x and CO to the atmosphere from coke production and fuel burning.	7. • Control by scrubbing with alkaline resolutions. • Analysis of raw materials during feasibility stage of project planning can determine existing levels of sulfur to properly design emission control equipment. • Strip, recycle and reuse carbon monoxide.

Table 10.14. Iron and Steel Manufacturing (continued)

Potential Negative Impacts	Mitigating Measures
Direct: Plant Operation (continued)	
8. Accidental release of potentially hazardous solvents, acidic and alkaline materials.	8. • Maintain storage and disposal areas to prevent accidental release. • Provide spill mitigation equipment, double wall tanks and/or diking of storage tanks.
9. Surface runoff of constituents, raw materials, coal, coal breeze and other substances frequently stored in piles on the facility grounds can pollute surface waters or percolate to ground waters.	9. • Rainwater percolation and runoff from solid materials, fuel and waste piles can be controlled by covering and/or containment to prevent percolation and runoff to ground and surface waters. • Diked areas should be of sufficient size to contain an average 24 hour rainfall.
Indirect	
10. • Occupational health effects on workers due to fugitive dust, materials handling, noise or other process operations. • Accidents occur at higher than normal frequency because of level of skill or labor.	10. • Facility should implement a Safety and Health Program designed to: • identify, evaluate, monitor, and control safety and health hazards • provide safety training
11. Regional solid waste problem exacerbated by inadequate on-site storage or lack of ultimate disposal facilities.	11. Plan for adequate on-site disposal areas, assuming screening for hazardous characteristics of the leachate is known.

Table 10.14. Iron and Steel Manufacturing (continued)

Potential Negative Impacts	Mitigating Measures
Indirect (continued)	
12. Transit patterns disrupted, noise congestions created, and pedestrian hazards aggravated by heavy trucks transporting raw materials and fuel to/from facility.	12. • Site selection can mitigate some of these problems. • Special transportation sector studies should be prepared during project feasibility to select best routes to reduce impacts. • Transport regulation and development of emergency contingency plans to minimize risk of accidents.

NONFERROUS METALS

1. The metals that will be discussed in these guidelines are aluminum, ferroalloys, copper, lead, zinc and nickel. There are many more nonferrous metals; but because they are either produced in small quantities, in highly specialized processes, or as by-products of other operations, the Bank rarely participates in projects designed for their production.

Aluminum

2. Aluminum is produced from the ore bauxite, a hydrated aluminum oxide. The bauxite ore first has to be purified from other elements by dissolving the alumina with a strong caustic soda solution. The residue is filtered off (red mud) and reworked for alumina. The final residue is discarded. Pure alumina is separated after crystallization, thickening, filtering, and calcining. The alumina is then electrolytically reduced, alloyed, and cast into ingots.

3. Secondary aluminum production has aluminum scrap and returned metal as feed. The scrap and recycled aluminum are smelted in a furnace, adding fluxing agents; they are then alloyed, demagged (magnesium removal), and degassed with chlorine and skimmed before casted into ingots.

Ferroalloys

4. The principal step in ferroalloy production is the reduction and smelting of mixed oxides in an electric furnace. The carbon in coke, which is usually added to the feed, removes the oxygen as carbon monoxide gas. The non-reducible oxides report in the slag and the reducible metals form an alloy. The molten slag and alloy are periodically tapped from the bottom of the furnace. The type of alloy -- ferrochrome, ferromanganese, ferronickel, ferrosilicon, ferrovanadium, ferroniobium, etc. -- depends on the composition of the ore feed to the furnace. The slag and alloy are cooled and separated. The alloy is broken, crushed, and screened to size for market.

5. The main environmental concerns in ferroalloy production are the carbon monoxide gas and the large quantity of ultra-fine dusts created in the electric furnace. In the past, furnaces have been open; however, modern plants use closed furnaces which improves the efficiency and greatly assists in controlling the gases and fumes produced by high-temperature operations. The gases are cleaned by cyclones, bag houses and scrubbers. The fine dusts are agglomerated and returned to the furnace. The carbon monoxide is used as fuel to preheat the feed or to fire boilers.

6. Molten slag and, less frequently, the alloy are sometimes granulated in a jet of water. This yields a liquid effluent and a fine solid slag which must be impounded, as both have some potential for environmental degradation. The furnaces are cooled with water, which produces another effluent stream.

Copper and Nickel Sulfide Smelting

7. The major portions of the world's copper and nickel are produced by the pyrometallurgical process of sulfide smelting. The principal step in this process is the melting and gravity separation of the low-density molten oxide slag from the higher density molten matte, which is a mixture of metal sulfides.

8. A roasting step which is used to adjust the sulfur and iron content of the furnace matte is usually found upstream of this smelting or melting step. In roasting, the feed is heated and reacted with air. The unwanted sulfur departs as sulfur oxide and the iron (which is usually present as a sulfide) as iron oxide which, in smelting, will go into the slag. An environmental concern in roasting is the presence, in many ores, of such impurities as arsenic, antimony, and cadmium. Their oxides tend to vaporize and later condense as dusts in the off-gas.

9. Converting of the matte follows smelting. Air, sometimes enriched in oxygen, is blown into the molten matte to remove sulfur and iron. The product is blister copper (an impure metallic copper) or iron-free sulfides, both of which require further processing. Converting is a high-temperature, high-gas-volume operation, which tends to eliminate impurities from the matte (e.g., oxides of lead, arsenic, and cadmium).

10. The equipment used in each of the steps described above has lately gone through many changes, motivated either by economics or environmental protection. Their net effects have been to decrease fuel consumption and yield a lower volume of gas with higher sulfur dioxide content. The latter improvement facilitates dust removal and recovery of sulfur as sulfuric acid or liquid sulfur dioxide.

Lead

11. Smelting of lead ores and concentrates has typically involved sintering to remove the sulfur, oxidize the lead and agglomerate the fine material, followed by reduction melting in a blast furnace. In recent years, the People's Republic of China and Canada have adopted a direct smelting process in which lead sulfide concentrate is fed to one end of a molten bath, where injected oxygen removes the sulfur, while coal or gaseous reductant are injected at the other end to reduce lead oxides out of the slag that is formed. Slag is removed from one end and crude metal from the other. The crude lead may then be electrorefined.

Zinc

12. Sulfide minerals are the primary sources of zinc. Two routes are use to extract the metal: one is a combination of pyrometallurgy, hydrometallurgy, and electrometallurgy; and the other is a straight pyrometallurgical process. Both start by converting the sulfide to an oxide. In the pyrometallurgical process, zinc oxide sinter cake is fed to a blast furnace. The zinc metal vaporizes and is condensed as molten zinc from the off-gases. In the hydrometallurgical step, the zinc oxide is dissolved with sulfuric acid, the solution is purified, and the zinc is recovered by electrowinning (a plating process). Jarosite, an iron sulfate, is a solid waste product from the purification step; and the electrowinning operation has a tendency to produce a fine acid mist.

Potential Environmental Impacts

13. The principal environmental impacts from the production of **aluminum**, starting with the processing of the mined ore, are the disposal of red mud (a mixture of clays and highly corrosive caustic

soda), emissions from fuel burning, emissions from the aluminum electrolysis process, and waste liquid and slurry streams. The red mud can degrade receiving waters and groundwater.

14. Emissions from the electrolysis plant consist of hydrofluoride, an extremely corrosive and hazardous gas, and carbon monoxide. Magnesium and gases from the demagging and degasing processes, respectively, contain chlorine and must be scrubbed. The scrub liquor most then be neutralized.

15. **Ferroalloy** production generates large amounts of fine dust and fine coke (coke breeze). The electrofurnaces produce large volumes of toxic gases, including carbon monoxide and some arsenic compounds. If not usable for other purposes, the slag has to be disposed of. Off-gases can be cleaned of dust through cyclones and baghouses, or further purified by scrubbing. The recovered dust can be recycled through a pelletizing plant. Scrubbing delivers an effluent which cannot be discharged without treatment.

16. The environmental impacts from the production of **nickel** depend on the process. The direct electrometallurgical production of ferronickel will produce large quantities of particulates and carbon monoxide gas, and small emissions of sulfur containing gases. The pyrometallurgical processes with matte production and emit gases that are rich in particulates and toxic gases originating from the roasters, smelters and converters, as well as from power generation, which is often part of the production facilities.

17. Gases can contain sulfur dioxide, nitrogen oxides, carbon monoxide, and hydrogen sulfide. Liquid effluents are generated by gas scrubbing and water cooling of converter matte and slag, furnace matte, reduction kilns, etc. Solid waste consists of slag, settled solids from cooling pits, and sludges from waste treatment. If the carbonyl process is applied, nickel-carbonyl, a highly poisonous gas, is formed as an intermediate.

18. **Copper** smelting and refining gases contain sulfur dioxide and particulates. The sulfur dioxide should be recovered and processed into sulfuric acid. Liquid effluents originate from acid plant blowdown, contact cooling, and slag granulation. Refining plant effluents contain waste electrolyte and cathode wash, fine slag, and anode mud.

19. Secondary copper production generates effluent from slag milling, smelter air pollution control, contact cooling, electrolyte and slag granulation. Solid waste is mainly from air scrubbers, cyclones and precipitators, furnace slag, and from secondary copper production as scrap or pretreatment waste.

20. Air pollutants from **lead** processing include particulate matter, sulfur dioxide, arsenic, antimony, and cadmium in the sintering plant. The highly concentrated sulfur dioxide stream from the blast furnace plant should be recovered in a sulfuric acid plant. Particulates are high in lead and should be removed in baghouses or scrubbers.

21. Liquid effluents, which may contain toxic metals, originate from the sintering plant scrubbers, acid plant blowdown, and other scrubbers in the plant. Slag granulation is another source of effluents. The effluents contain lead, zinc, copper, and cadmium. Solid wastes originate from cyclones bagfilters, etc., and for the most part can be reused in the plant.

22. Secondary lead plants produce effluents containing battery acid from cracked battery washers and from scrubbing equipment for air pollution control. The battery acid is contaminated with lead, antimony, cadmium, arsenic and zinc, and should be kept separate from other wastes and not discharged.

23. Emissions in the pyrometallurgical **zinc** process contain sulfur dioxide, arsenic, lead, and cadmium. The sulfur dioxide is recovered for sulfuric acid production. Carbon monoxide is an important component from the reduction furnace off-gases. Uncondensed zinc fumes are scrubbed and returned to the refining process. The electrometallurgical zinc process has the same type of air emissions, with the occasional addition of mercury (removed in a scrubber). Effluents from scrubbers, acid plant blowdown, and from leaching units may contain the same elements as the air emissions.

24. Solid wastes contain significant quantities of other metals and are normally sold to other processors. Cadmium, however, is an exception; its recovery is nearly always practiced at the zinc production site. (For more information on the environmental impacts from nonferrous metal production, see Table 10.15 at the end of this section.)

Special Issues

Air Emissions

25. The production of **aluminum** from alumina by electrolysis causes air emissions of fluorine which contain gases that can be very harmful for the environment and for human health. These emissions require careful monitoring. These gases are normally dry scrubbed with alumina powder which eliminates most of the fluorine. The remainder has to be removed by wet alkaline scrubbing.

26. Substantial particulate emissions can occur in the production of **ferrochrome** and **ferromanganese**. These can be minimized in the design phase by choice of furnace (open, semi-open, or closed) and the inclusion of a pelletizing unit to return fines to the process.

27. In most plants, sulfur dioxide gas from roasting of sulfide ores is recovered, cleaned, and used as feedstock for producing sulfuric acid. The gas cleaning produces effluents with arsenic, selenium, and other toxic metal salts which cannot be discharged into natural water streams, but have to be treated to remove these elements.

Effluents

28. In general, water effluents need not be a special issue if properly managed and monitored. All particulates should be settled and removed, and as far as possible, water should be recirculated in the process, if necessary after treatment. No discharge of water containing metal ions (metal salts) from the copper, chromium, manganese, nickel, zinc and lead processes should be allowed in concentrations over those indicated in the World Bank's Environmental Guidelines.

29. Any spent acid used for leaching or other treatment cannot be discharged into any natural water, but should be neutralized or reprocessed. If neutralized, discharge can only take place if the concentration of harmful metals and other components is below the limits officially allowed.

Solid Waste

30. In **aluminum** production, a great amount of red mud is generated which has to be disposed of. This waste material cannot be discharged into natural water streams, but has to be stored on land in such a way that no runoff water or leachate can contaminate streams or groundwater. In general, land impoundment in a lined diked area is the method most recommended and most frequently used in Bank-supported projects. Water from settling ponds and impoundment areas can be returned to the process after treatment. Ultimately, stabilization and revegetation are desirable.

31. The solid waste from production of most other nonferrous metals contains reusable materials; however, recycling should be a consideration in developing disposal measures. Sludges, if not sold for reprocessing, must be stored under controlled conditions to prevent leaching to groundwater or runoff to surface waters. Sludge from **lead** plants is a particular problem, since it may carry large concentrations of toxic metals.

Waste Minimization

32. In Bank-assisted projects, process water should be recycled in the processes. Solid waste can usually be sold to other processors for recovery of usable materials or, if it is innocuous, used for other purposes under carefully controlled conditions (like the use of red mud for seashore reclamation). However, if solid waste is to be sold or transferred to contractors either for further processing or for landfill, the project should prescribe strictly controlled conditions.

Safety in Handling Hot Metal

33. In all operations involving molten metal, there is the danger of explosion caused by contact with water. The mechanics of this explosion are not well understood. Flooding the metal with water, as in granulating matte, is safe, whereas small amounts of water on molten metal can be deadly.

Project Alternatives

34. Although a variety of alternatives exist for project planning and execution, the types of nonferrous manufacturing facilities suitable to the project are limited by available technologies and raw materials.

Site Selection

35. General issues to consider in industrial plant siting are discussed in the section on "Plant Siting and Industrial Estate Development." The nature of nonferrous metal production is such that impacts on water quality and land through solid waste disposal from the production processes warrant special attention in evaluating alternative sites. Receiving waters with substandard quality or insufficient flow to accept even well-treated effluents are inappropriate.

36. If mining and production are at the same site or very near to each other, the total impact on the environment from both operations should be evaluated. There could be a positive result in that old mining sites could be used to deposit solid waste materials under strictly controlled conditions.

Manufacturing Process

37. Nonferrous metals production processes vary with the metals to be produced and with the raw materials that are used. Although often not a point of consideration in a particular project, it should be pointed out that, in general, any in-country recycling possibilities of scrap metal should be considered and exhausted before new metal production facilities are developed. This will not only be beneficial from an environment point of view, but could also save the country high energy consumption cost in production as well as in mining.

38. For **aluminum** production, it is important to check whether the latest technological developments have been included that can have a beneficial effect on waste management, such as fluid beds for waste heat recovery from aluminum melt furnaces.

39. For the production of **nickel**, **copper**, and **zinc** from sulfide ores, two different process routes are often available: a pyrometallurgical and a hydrometallurgical one. The process selection is based on many different considerations, from inherent ore properties to such non-metallurgical factors as geographical location, water and power availability, and market requirements. The advantage of hydrometallurgy is that it is well suited to low-grade and more complex ores. This is important because the world's high grade ores are becoming depleted. And it can often be used for small ore deposits using relatively small process plants. However, the assertion that a hydrometallurgical process is preferable to a pyrometallurgical one for environmental reasons is not necessarily defensible: the situation is not clear-cut and must be evaluated separately for each project.

Air Pollution Controls

40. Air pollution control is required in Bank-assisted projects. Alternatives to be evaluated include:

- design of process and choice of equipment, electrostatic precipitators, flue gas (wet or dry)
- electrostatic precipitators
- flue gas scrubbers (wet or dry)
- high efficiency cyclones
- baghouse filters
- sulfur dioxide separation and utilization for sulfuric acid
- carbon monoxide separation and utilization for heating

Water Quality Controls

41. Water pollution control alternatives include:

- wastewater reuse
- solar evaporation

- precipitation
- solar evaporation
- flocculation, sedimentation, clarification, and filtration
- ion exchange, membrane filtration, reverse osmosis
- neutralization (active pH control)
- biological treatment, if required

Management and Training

42. The potential negative impacts on air and water quality from all nonferrous metallurgical processes necessitate institutional support for efficient management of pollution control and waste reduction. The plant staff should include a plant engineer trained in air and water pollution control and the monitoring technologies being used especially in nonferrous industries. Manufacturers will sometimes supply the necessary equipment operations and maintenance training, if requested.

43. Standard operating procedures and predictive maintenance should be established for the plant and enforced by management. They should include pollution control equipment operation, air and water quality monitoring requirements, and instructions for notification and shutdown or other responses to pollution equipment failure.

44. Plant health and safety rules should be established and enforced. They should include:

- Provisions to prevent and respond to accidental gas leaks and acid spills.

- Procedures to keep exposure to toxic gases and air borne particulates below limits established by country or World Bank regulations.

- A program of routine medical examinations.

- Ongoing training in plant health and safety and in good environmental housekeeping practices.

- Emergency procedures, regular drills, and a plan of action in case of a major spill, leak, explosion or fire.

(For further discussion and guidance, see the section on "Industrial Hazard Management" and the World Bank's Occupational Health and Safety Guidelines.)

45. Emissions and effluent standards should be set for the plant, based on national regulations where they exist, or on World Bank guidelines. Government agencies charged with the operation of pollution control equipment, monitoring air and water quality, enforcing standards, and overseeing waste disposal should have the necessary equipment and specialized training that are required. These activities should be financed by the project. The environmental assessment should also include an evaluation of local capabilities in these areas and recommend appropriate assistance to be included in the project.

Monitoring

46. Monitoring plans are necessary for plant/site specific. In general, however, metallurgical nonferrous works should be monitored for the following:

- flue gas opacity
- emissions of particulates, sulfur dioxide, hydrogen fluoride, hydrogen sulfide, chlorine, ammonia, nitrogen oxides, where applicable
- process parameters which indicate the operation of air pollution control equipment, such as flue gas temperature
- work space air quality as applicable to plant type and process as possible
- ambient air quality in vicinity of plant for applicable pollutants
- receiving water quality downstream for dissolved oxygen, pH, total suspended solids, cyanide, free chlorine and relevant toxic metals
- liquid waste streams from plants for pH (continuous), total suspended solids, total dissolved solids, and where applicable, for cyanide, hydrogen sulfide, hydrogen fluoride, sulfuric acid, caustic soda, toxic metal ions, radioactivity, pH, BOD_5, oil and grease
- permitted storm water discharges from plants and storage areas for above mentioned pollutants
- working areas of all plants for ambient noise levels
- storage piles of waste materials, pond piles and sludge storage in diked areas for runoff, infiltration and leachate
- inspection for adherence to safety and pollution control procedures

Table 10.15. Nonferrous Metals

Potential Negative Impacts	Mitigating Measures
Direct: Site Selection	
1. Siting of plant on/near sensitive habitats such as mangroves, estuaries, wetlands, coral reefs.	1. • Locate plant in industrially zoned area, if possible, to minimize or concentrate the stress on local environmental services and to facilitate the monitoring of discharges. • Involve natural resource agencies in site selection process to review alternatives.
2. Siting along water courses causing their eventual degradation.	2. • Site selection process should examine alternatives which minimize environmental effects and do not preclude beneficial use of water bodies. • Plants with liquid discharges should only be located on a watercourse having adequate capacity to assimilate waste in treated effluent.
3. Siting can cause serious air pollution problems for local area.	3. • Locate plant at elevation above local topography in an area not subject to air inversions, and where prevailing winds are towards relatively unpopulated areas.
4. Siting can aggravate solid waste problems in an area.	4. • Site selection should evaluate the location according to the following guidelines: • proximity to suitable disposal site • plot size sufficient for landfill or disposal on-site • convenient for public/private contractors to collect and haul solid wastes for final disposal • reuse or recycle materials to reduce waste volumes

Table 10.15. Nonferrous Metals (continued)

Potential Negative Impacts	Mitigating Measures
Direct: Plant Operation	
5. • Water pollution from discharge of liquid effluents and process cooling water or runoff from waste piles. • Plant: metals, oil and grease, ammonia nitrogen • Materials storage piles runoff: TSS, pH, metals	5. • Laboratory analysis of liquid effluent should include: metals, TSS, oil and grease, ammonia nitrogen, pH, and in-situ temperature monitoring. All Plants • No cooling water discharge. If recycling not feasible, discharge cooling water provided receiving water temperature does not rise >3°C. • Maintain pH level of effluent discharge between 6.0 and 9.0. • Control effluent to specified limitations in Bank or other guidelines (e.g., EPA 40 CFR 421) for specific process. Material Storage Piles/Solid Waste Disposal Areas • Minimize stormwater allowed to percolate through materials and runoff in uncontrolled fashion. • Line open storage areas.
6. Particulate emissions to the atmosphere from all plant operations.	6. Control particulates by fabric filter collectors or electrostatic precipitators.
7. Gaseous emission to the atmosphere from metals processing and fuel burning.	7. • Control by scrubbing with alkaline solutions. • Analysis of raw materials during feasibility stage of project planning can determine existing levels of sulfur to properly design emission control equipment.

Table 10.15. Nonferrous Metals (continued)

Potential Negative Impacts	Mitigating Measures

Direct: Plant Operation (continued)

8. Accidental release of potentially hazardous solvents, acidic and alkaline materials.

    8.
    - Maintain storage and disposal areas to prevent accidental release.
    - Provide spill mitigation equipment, double wall tanks and/or diking of storage tanks.

9. Surface runoff of constituents, raw materials, and other substances frequently stored in piles on the facility grounds can pollute surface waters or percolate to ground waters.

    9.
    - Rainwater percolation and runoff from solid materials, fuel and waste piles can be controlled by covering and/or containment to prevent percolation and runoff to ground and surface waters.
    - Diked areas should be of sufficient size to contain an average 24 hour rainfall.

Indirect

10.
    - Occupational health effects on workers due to fugitive dust, materials handling, noise or other process operations.
    - Accidents occur at higher than normal frequency because of level of skill or labor.

    10.
    - Facility should implement a Safety and Health Program designed to:
        - identify, evaluate, monitor, and control safety and health hazards
        - provide safety training

11. Regional solid waste problem exacerbated by inadequate on-site storage or lack of ultimate disposal facilities.

    11. Plan for adequate on-site disposal areas, assuming screening for hazardous characteristics of the leachate is known.

12. Transit patterns disrupted, noise congestions created, and pedestrian hazards aggravated by heavy trucks transporting raw materials and fuel to/from facility.

    12.
    - Site selection can mitigate some of these problems.
    - Special transportation sector studies should be prepared during project feasibility to select best routes to reduce impacts.
    - Transport regulation and development of emergency contingency plans to minimize risk of accidents.

Table 10.15. Nonferrous Metals (continued)

Potential Negative Impacts	Mitigating Measures
Indirect (continued)	
13. Mining of ore and coal locally for metals manufacturing can create conflicts with other industries (coal for utilities), and aggravate erosion/sedimentation of water courses by uncontrolled or unrestricted operations.	13. • Plan for coal resource usage to fit availability and impose restrictions on manner of mining. • Coordination with responsible agency-in-charge to examine site reclamation options once facility decommissioned.
14. Metals processing may require significant amounts of electricity which may result in conflicts with other industrial users.	14. • Operate metals processing operations at hours when other power consuming industries are not operating. • Increase electrical power generation capabilities.

PETROLEUM REFINING

1. This category includes the production of a wide range of petroleum and chemical products, fuels, lubricants, bitumen, and chemical feed stocks from crude oil. Petroleum refining is completed through the following steps: (a) separation of oil into fractions according to boiling range and eventual products, (b) conversion of compounds by splitting, rearranging, or recombining component molecules, (c) treatment to remove contaminants, such as sulfur and (d) blending of products with additives to meet product specifications.

Potential Environmental Impacts

2. The environmental impacts of petroleum refining result primarily from gaseous emissions, wastewater discharges, solid wastes, noise, odor, and visual or aesthetic effects.

3. Atmospheric emissions are the most significant causes of adverse environmental impacts from refineries. Most important are particulates, hydrocarbons, carbon monoxide, sulfur oxides, and nitrogen oxides. They emanate from various sources including the catalytic cracking unit, sulfur recovery processes, heaters, vents, flares and product or raw material storage. Pump seals and valves can be fugitive emission sources. The combination of emissions can produce obnoxious odors affecting large areas in the neighborhood of the refinery.

4. Large quantities of water are used in petroleum refining for washing unwanted material from the process stream, for cooling and steam production, and in the reaction processes. The major pollutants present in petroleum refinery wastewater discharges are oil and grease, ammonia, phenolic compounds, sulfides, organic acids, and chromium and other metals. These pollutants may be expressed in terms of biochemical oxygen demand (BOD_5), chemical oxygen demand (COD), and total organic carbon (TOC). In addition, there is potential for serious surface water, soil and groundwater contamination from leaks or spills of raw materials or products. Cooling water blowdown, flushing and cleaning water, stormwater runoff and percolation from tank farms, pipe racks, product loading areas, and processing areas can also cause surface water and groundwater degradation.

5. Refineries generate large volumes of solid wastes; chief among them are catalytic fines from cracking units, coke fines, iron sulfides, filtering media, and various sludges (from tank cleaning, oil and water separators, and wastewater treatment systems).

6. Petroleum refining can be a noisy operation. Sources of noise include high speed compressors, control valves, piping systems, turbines and motors, flares, air cooled heat exchangers, fans, cooling towers and vents. Typical noise levels range from 60 to 110 dB at a distance of one meter from the source. (See Table 10.16 at the end of this section for other examples of adverse environmental impacts produced from petroleum refineries and recommended measures to avoid or mitigate them.)

Special Issues

Risks of Accidental Releases

7. A major release or spill of raw materials, products or wastes can be environmentally catastrophic, especially to marine or aquatic ecosystems. Groundwater is particularly vulnerable to contamination by undetected leaks from tanks or pipelines. Refineries should be sited away from areas prone to natural disaster (earthquakes, storm tides, floods, adverse meteorological conditions, etc.) and away from sensitive resources that cannot be protected in the event of a major release. Designs for storage and transfer facilities should include features to contain spills. Pipelines should be equipped with alarms and automatic shutoff valves to provide fast response to breaks. Plant operating procedures should include frequent inspections of tanks and pipelines for leakage.

8. Training in safety and spill response should be routine for personnel involved in the transportation of raw materials and products. A spill response plan should be developed with local government authorities and hospitals as part of the project, including the notification of officials and affected parties (e.g., downstream water users, fishing fleets, ports and marinas, tourist areas), provisions for assigning responsibility for containment and clean-up, evacuation procedures, medical attention, and advance acquisition of equipment and supplies.

Explosion and Fire Hazards

9. Raw materials and petroleum products are typically combustible or explosive. These hazards should be considered in siting refineries; to reduce the danger posed by them, designs and procedures should be implemented at each facility. In addition, emergency fire equipment should be provided at a refinery. The capacity of the surrounding communities to respond to disasters should be evaluated and strengthened, if necessary. For more details, see the section on "Industrial Hazard Management."

Waste Minimization, Recycling, and Reuse

10. There are two types of in-plant measures that can greatly reduce the volume of refinery effluent. The first is to reuse water from one process in another; for example, to use blowdown from high-pressure boilers as feedwater for low pressure boilers or treated effluent as make-up water wherever possible. The second approach is to design systems that recycle water repeatedly for the same purpose. Examples are employing cooling towers or using steam condensate as boiler feedwater.

11. Good housekeeping, combined with good work practices, will further reduce waste flows. Examples are minimizing waste when sampling product lines, using vacuum trucks or dry cleaning methods for spills, applying sound inspection and maintenance practices to minimize leakage, and segregating waste streams having special characteristics for disposal (e.g., spent cleaning solution).

Project Alternatives

Site Selection

12. General issues to consider for industrial plant siting are discussed in the "Plant Siting and Industrial Estate Development" section. The nature of petroleum refining is such that potential impacts of air quality, water resources, and aesthetics warrant special attention in evaluating alternative sites. Refinery siting requirements include:

- water supplies of adequate quality and quantity to supply refinery needs and assimilate treated effluent without impairment of desired uses or receiving waters;

- sufficient land area to provide logical and unrestricted placement of facilities for storage of raw materials, manufacturing, maintenance and waste disposal, and to allow for future expansion;

- compatibility of adjacent land uses, i.e., adequate distance from residential, commercial, institutional, recreational, and tourist sites to avoid air quality, odor and noise impacts, as well as explosions and fire hazards;

- appropriate topography to reduce impacts of adverse meteorological conditions;

- low risk of damage from natural hazards;

- avoidance of groundwater recharge areas; and

- adequate distance from cultural properties that can be damaged by emissions from refineries.

Material Transport

13. Most major oil spills result from transportation accidents. Each mode of transporting raw materials to and products from a refinery has its own risks of accidental release during transfer or hauling. The level of risk varies to some extent with geographical setting and state of the country infrastructure. The risk of accidents and the sensitivity and importance of the ecological and sociocultural resources that might be damaged can be weighed against the costs of alternative modes of transportation and their own environmental impacts in deciding which alternatives to employ at a given refinery site. There are cases where the potential impacts of siting a refinery at a particular location can only be reduced to acceptable levels by selection of one transportation method; for example, above-ground or underground pipelines over inland routes to and from a plant would be superior to tanker, barge, rail or truck transport in a sensitive coastal area with important wetlands.

Process Modification

14. In most cases, process modifications that are environmentally beneficial and applicable to both existing and new installations could include:

- substitution of improved catalysts having longer life and requiring less frequent regeneration;

- substitution of air fan cooling for water cooling (to reduce blowdown discharges), and recirculating water system for once-through water system;

- maximization of hydrogen-addition processes and minimization of carbon-removal and chemical treatment processes to generate the lowest possible waste loadings; and

- maximum use of improved drying, sweetening, and finishing procedures to minimize volumes of spent caustics, filter solids, and other materials requiring special provisions for disposal.

Management and Training

15. The potential impacts of petroleum refining on air, water, and soil necessitate institutional support for efficient conduct and supervision of materials handling, pollution control and waste reduction. Facility personnel should be trained in air and water pollution control technologies being employed. Equipment manufacturers will frequently supply the necessary training in equipment operations and maintenance. Standard operating procedures should be established for the refinery and enforced by management. They should include pollution control equipment operation, air and water quality monitoring requirements, special steps to avoid flare emissions from steam injections, instructions for operators to prevent malodorous emissions, and directives for notification of proper authorities in the event of accidental release of pollutants. Toxic and hazardous substance handling and management should be improved by detectors, alarms, etc., and special training of operating personnel.

16. Emergency procedures are necessary to provide for rapid and effective action in the event of accidents that pose serious threats to the environment or to the surrounding community, such as those caused by major spills, fires, or explosions. Local government officials, agencies, and community services (medical, firefighting, etc.) usually play key roles in these types of emergencies and should be included in the planning process. Periodic drills are important components of response plans (for more information, see the "Industrial Hazard Management" section).

17. Plant health and safety rules should be established, including procedures to keep exposure to noise and toxic substances below accepted limits, a program of routine medical examinations and monitoring of clinical medical records, and ongoing training in plant health and safety and good environmental housekeeping practices. (See the "Industrial Hazard Management" section.)

18. Emission and effluent standards should be set for the plant based on national regulations where they exist, or on World Bank guidelines where they do not. Government agencies should have the necessary equipment, authority, and appropriate training required for monitoring the operation of pollution control equipment, enforcing standards, and responding to emergencies. The environmental assessment should include an evaluation of local capabilities in these areas and recommend appropriate elements of assistance to be included in the project.

Monitoring

19. Monitoring plans are necessary for plant- and site-specific. In general, however, monitoring at a refinery includes:

- flue gas opacity (continuous)
- periodic stack testing for particulate matter, sulfur oxides, nitrogen oxides (for fuel burning units and fluid catalytic cracking unit), hydrogen sulfide (for hydro-desulfurization and sulfur recovery units)
- ground level concentrations (GLC) at various distances from the site
- wastewater oil content (continuous)
- local meteorological station for year-round tracking of weather conditions
- periodic sampling of wastewaters (24 hour composite sample) for biochemical oxygen demand (BOD_5), chemical oxygen demand (COD), total organic carbon (TOC), total suspended solids (TSS), oil and grease, phenolic compounds, ammonia nitrogen, sulfides, total chromium, pH, temperature and flow
- continuous monitoring of certain parameters to assure early detection of process upsets
- avoid excessive pollutant discharges -- e.g., total organic carbon (TOC) and flow
- installation of monitoring wells and periodic sampling of groundwater to provide early warning of contamination from spills and leaks

Table 10.16. Petroleum Refining

Potential Negative Impacts	Mitigating Measures
Direct: Site Selection	
1. Siting of refinery on/near sensitive habitats such as mangroves, estuaries, wetlands, coral reefs.	1. • Locate refinery in industrially zoned area, if possible, to minimize or concentrate the stress on local environmental services and to facilitate the monitoring of discharges. • Integrate site selection process with natural resource agencies to review alternatives.
2. Siting along water courses causing their eventual degradation.	2. • Site selection process should examine alternatives that minimize environmental effects and not preclude beneficial use of the water body using the following guidelines: • on a watercourse having adequate waste assimilative capacity • in an area where wastewater can be reused with minimal treatment for agricultural or industrial purposes • within a municipality that is able to accept the plant wastes in their sewage treatment system
3. Siting can cause serious air pollution problems for local area.	3. Locate refinery in an area not subject to air inversions or trapping of air pollution, and where prevailing winds are towards relatively unpopulated areas.
4. Siting can aggravate solid waste problems in an area.	4. For facilities producing large volume of waste, site selection should evaluate the location according to the following guidelines: • plot size sufficient to landfill or dispose on-site • proximity to suitable disposal site • convenient for public/private contractors to collect and haul solid wastes for final disposal

Table 10.16. Petroleum Refining (continued)

Potential Negative Impacts	Mitigating Measures
Direct: Plant Operation	
5. • Water pollution from discharge of liquid effluents and process cooling water or runoff from waste piles may contain: BOD, COD, TOC, oil and grease, ammonia, phenolic compounds, sulfides, and chromium.	5. • Control by wastewater reuse, at-source pretreatment and end-of-pipe control technology. (a) Major at-source pretreatment measures include: • stripping of sour waters • neutralization and oxidation of spent caustics (b) End-of-pipe control technology relies on a combination of flow equalization, physical-chemical methods (such as dissolved air flotation and sludge thickeners), and biological methods (such as activated sludge, aerated lagoons or trickling filters).
6. • Air pollution from refinery operations: (a) Storage vessels -- hydrocarbons (HC) (b) Refinery process gas -- hydrogen sulfide (H_2S) (c) Catalyst regenerators -- particulates, carbon monoxide (CO) (d) Accumulator vents -- HC (e) Pumps and compressors -- HC	6. • Source control measures to reduce air contaminants and odors: (a) vapor recovery systems, floating-roof tanks, pressure tanks, vapor balance, painting tanks white (b) ethanolamine absorption, sulfur recovery (c) cyclones-precipitator in-situ CO combustion, CO boiler, cyclones-water scrubber, multiple cyclones, electrostatic precipitator, bag filter (d) vapor recovery and vapor incineration (e) mechanical seals, vapor recovery, sealing glands by oil pressure, maintenance

Table 10.16. Petroleum Refining (continued)

Potential Negative Impacts	Mitigating Measures
Direct: Plant Operation (continued)	
(f) Vacuum jets -- HC	(f) vapor incineration
(g) Equipment valves -- HC	(g) inspection and maintenance
(h) Pressure relief valves -- HC	(h) vapor recovery, vapor incineration, rupture discs, inspection and maintenance
(i) Effluent waste disposal -- HC	(i) enclosure of separators, covering of sewer boxes, use of liquid seal, liquid seals on drains
(j) Bulk-loading facilities -- HC	(j) vapor collection with recovery or incineration, submerged or bottom loading
(k) Acid treating -- HC, sulfides, mercaptans	(k) continuous-type agitators with mechanical mixing, replacement with catalytic hydrogenation units, incineration of all vented gases, cessation of sludge burning
(l) Acid sludge storage and shipping -- HC	(l) same as (k)
(m) Spent-caustic handling -- sulfides, mercaptans	(m) Steam scrubbing, neutralization incineration, return system
(n) Sweetening processes -- HC	(n) steam stripping of spent doctor solution to hydrocarbon recovery before air regeneration, replacement of treating unit with other less objectionable units
(o) Sour-water treating -- ammonia (NH_3)	(o) use of sour-water oxidizers and gas incineration, conversion to ammonium sulfate

Table 10.16. Petroleum Refining (continued)

Potential Negative Impacts	Mitigating Measures

Direct: Plant Operation (continued)

(p) Mercaptan disposal

(q) Asphalt blowing -- HC

(r) Shutdowns, turnarounds -- HC

(s) Boilers and heaters -- SO_x, NO_x, particulates

(t) Sulfur recovery unit (Claus) -- SO_2

(u) Solvents (hydrocarbons, amines)

7. Noise Emissions

8. Accidental release (spills) of raw materials, products, potentially hazardous solvents, chemicals, acidic and alkaline materials.

(p) conversion to disulfides, adding to catalytic cracking charge stock; incineration, use of material in organic synthesis

(q) incineration, water scrubbing (non-recirculating type)

(r) depressurizing and purging to vapor recovery

(s) fuel hydro-desulfurization, flue gas desulfurization

(t) provide tail gas treatment; spare unit put into operation during main unit downtime

(u) provide closed circuit recovery units

7.
- Enclose noise emitting equipment/processes in structures to reduce potential for fugitive emissions.
- Employ other noise abatement procedures.

8.
- Inspect and maintain storage and disposal areas to prevent accidental release.
- Provide alarms, automatic shut-off valves, containment (bunding, enclosing) of accidental spills, spill mitigation equipment and emergency response plans.

Table 10.16. Petroleum Refining (continued)

Potential Negative Impacts	Mitigating Measures

Direct: Plant Operation (continued)

9. Surface runoff of constituents, raw materials, processing facilities and transfer areas can pollute surface waters or percolate to ground waters.

9. • Rainwater percolation and runoff follow appropriate regulators for product or raw material transport can be controlled by covering and/or containment to prevent percolation and runoff to ground and surface waters.

   • Diked areas should be lined and of sufficient size to contain an average 24 hour rainfall.

Indirect

10. • Occupational health effects on workers due to fugitive dust, materials handling, noise or other process operations.

    • Accidents occur at higher than normal frequency because of level of skill or labor.

10. • Facility develops a Safety and Health Program designed to identify, evaluate, monitor, and control safety and health hazards at a specific level of detail, and to address the hazards to worker health and safety and procedures for employee protection, including any or all of the following:

    - site characterization and analysis
    - site control
    - training
    - medical surveillance and tracking of clinical records
    - engineering controls, work practices and personal protective equipment
    - monitoring
    - informational programs
    - handling raw and process materials
    - decontamination procedures
    - emergency response
    - illumination
    - sanitation at permanent and temporary facilities

Table 10.16. Petroleum Refining (continued)

Potential Negative Impacts	Mitigating Measures

Indirect (continued)

11. Regional solid waste problem exacerbated by inadequate on-site storage.

    11. Plan for adequate on-site disposal areas assuming screening for hazardous characteristics of the leachate is known.

12. Transit patterns disrupted, noise and congestion created, and pedestrian hazards aggravated by heavy trucks transporting raw materials to/from facility.

    12.
    - Site selection can mitigate some of these problems.
    - Special transportation sector studies should be prepared during project feasibility to select best routes to reduce impacts.
    - Transporter regulation and development of emergency contingency plans to minimize risk of accidents.

13. Potential for increased land/surface water degradation by pipeline transport of products or new materials.

    13.
    - Siting of pipeline should be such as to minimize environmental hazards.
    - Develop program for periodic pipeline surveillance.

PULP, PAPER, AND TIMBER PROCESSING

1. This category comprises all manufacturing projects involving the production of paper, such as newspaper or kraft paper, soft tissue paper, and paperboard.

2. The pulp and paper manufacturing can be divided into a two step process: a) pulping of a great variety of fibrous materials from wood or other plant fibers or, in growing volume, from recycled paper, and b) the production of paper products. Minor amounts of synthetic fibers are used for specialty papers.

3. The production of paper can be combined with the pulping (integrated papermills) or be separate, in which case the pulp is bought from pulpmills in the country or imported. In industrial countries, pulp mills seldom have a capacity of less than 500 tons of pulp per day. In developing countries, pulp mills can be as small as 50 tons per day.

4. The pulping processes can be mechanical, thermo-mechanical and chemi-thermo-mechanical, or chemical with either sulfite, kraft or kraft/sulfite processes. The kraft process is the dominant pulping process because of its versatility and flexibility. Some older plants use the sulfite process which was dominant until 1935, because at that time sulfite pulp was considerably cheaper and easier to produce than kraft pulp.

5. In an integrated papermill, the pulp slurry is directly piped to the paper machines. A non-integrated mill obtains the pulp mostly in dried form. The dried pulp is mixed with water before being fed to the papermill.

6. The primary raw material for pulping is wood, but other plant fibers are used also -- straw, bagasse, bamboo, papyrus, sisal, flax, jute, etc. Wastepaper is an increasingly important raw material, especially for the production of newsprint paper and certain tissues, writing paper, magazines and boxboard. De-inking is the only chemical treatment as most of the recycled paper is mechanically pulped.

7. The pulp mills are often close to their resource base, namely, the forests. Good forest management is important to ensure a steady and sustained supply of wood and also because the logging of trees is one of the most difficult and hazardous operations in the paper industry.

Potential Environmental Impacts

8. Pulp and paper mills built prior to 1970 were planned in an economic and social climate different from that of today. Increased costs of construction, raw materials and energy, and a greatly increased environmental awareness have radically affected the design and operating philosophies of the pulp and paper industry.

9. Adverse impacts on the environment (e.g., natural and tropical forests degradation) by the development of wood resources to feed the manufacturing mills has led to serious problems in land erosion, watershed management, and loss or degradation of forest habitat. Crop pests can expand

unchecked when forests are converted to monoculture. This procedure requires use of pesticides and/or herbicides, which have toxicologic effects on beneficial organisms. (See Table 10.17 at the end of this section for further discussion.)

10. Wood harvesting can have severe environmental and health impacts, as well. One of the most hazardous occupations is logging and, if not properly supervised, it can affect soil fertility and promote soil erosion that causes increased turbidity in streams, lakes, and estuaries. Also, chemical changes in the waterways can occur if large quantities of waste wood, bark, and litter are allowed to decompose in them.

11. Pulp and papermills use great quantities of water in the preparation of wood by wet debarking. Although wet debarking is physically the most efficient way to remove bark with less wood loss and less dirt, the higher cost for effluent control of the wastewater and the lower heating value of wet bark are one of the primary reasons today to convert from wet to dry debarking. The most important parameters for controlling pollution in wet debarking are TSS, BOD_5, pH, color, and toxicity.

12. The use of non-wood fibers like straw, bamboo, and bagasse eliminates debarking but requires pretreatment through washing to remove dirt, grit, and pith in bagasse. All of the various pulping processes generate substantial effluent streams that have to be treated and, as far as possible, recycled; most processes cause air pollution.

Kraft and Soda Pulping

13. Liquid effluents from the kraft and soda pulping consist of spent liquor and contaminated condensates. The condensate is often processed into turpentine, as a valuable byproduct, and has to be treated before discharge by air stripping or steam stripping. Although steam stripping is more expensive, it is often preferred above air stripping because of much smaller gas volumes to be handled. Odor control can be a great problem. Major components in the condensate are toxic, reduced sulfur compounds, and methanol. An additional operation, bleaching, can increase the toxicity of the condensate when chlorine is used. Oxidative bleaching, on the other hand, will reduce the toxicity and color of the effluent water.

14. The "black liquor" that is produced from pulp washing has to be concentrated by evaporation and burned afterwards. For this process, a multiple effect evaporator with steam should be used. Because of considerable hydrogen sulfide emissions, the older method of direct evaporation with flue gases should be discouraged. Sulfur is recovered as sodium disulfide; it is converted separately in caustic soda for recycling mainly as sodium bicarbonate.

15. Gaseous emissions from the kraft and soda process consist of sulfur compounds, organic compounds, sulfur dioxide, and nitrogen oxides. The sulfur compounds, especially, can cause severe odor problems. The gases have to be collected and scrubbed carefully. The recovery boiler or furnace where spent liquor is incinerated can be a major source of particulates emission, as is the smelt dissolving tank and the lime kiln.

Sulfite Pumping

16. Sulfite pulping produces effluents and emissions different in composition from those of the kraft process. The spent liquor is evaporated and burned in a recovery furnace and the sulfur dioxide that is formed is absorbed in a chemical recovery system. Depending on the basic sulfite solution used, sodium and magnesium can be recovered and recycled; however, calcium and ammonia pose problems. Ammonia is oxidized in the furnace into nitrogen and nitrogen oxides.

17. The toxicity of effluent water should be controlled carefully. Air pollution in the sulfite process is rather different from that in the kraft process. Sulfur dioxide is a major pollutant and requires careful designing of the acid preparation system and the digester gas relief system to prevent air pollution.

Mechanical and Thermo-mechanical Pulping

18. Mechanical and thermo-mechanical pulping processes primarily use softwood. These processes are the simplest method for producing wood pulp, and the total amount of waste material is substantially smaller than from the chemical processes. Mechanical pulping converts 90 to 95 percent of the wood into pulp, compared to about 50 percent from the kraft process. Air pollution is minimal and water pollution depends mainly on the type of wood used and consists of carbohydrates, lignin, extractives, acetic acid, formic acid, methanol, and ash. Toxicity and BOD are caused by watersoluble solids, such as certain carbohydrates, extractives, and inorganic solutions from cell contents and rotting processes.

Paper Making

19. Paper making requires large quantities of water, most of which can be recycled after treatment. The characteristics of the effluents vary from mill to mill depending on the degree of water recycling, grade of paper produced, size of the mill, and raw material used. Pollutants will consist of suspended solids and dissolved substances from the wood fibers and from the additives used in the paper production.

Special Issues

Air Pollution

20. The main air pollution problems in the pulping plants are malodorous sulfur compounds with extremely low detection threshold levels (1 and 10 parts per billion [ppb]). These gases are produced primarily in kraft pulping plants, from digester blow and relief valves, vacuum washer hood and seal tank vents, multiple effect evaporation hotwell vents, recovery furnace flues, smelt dissolving tanks, slaker vents, black liquor oxidation tank vents, and wastewater treatment operations. Preventing releases of sulfur gases requires a design that collects all off-gases, including the incidental releases, and an adequate incineration system with scrubbing of the exhaust gases.

21. Incidental chlorine emissions through tank vents, wash filters, and sewers in pulp bleaching operations are another point of concern in plant design.

Effluent Systems

22. The main problems here are the high BOD and COD levels in water discharged from the plant and the black liquor effluent. All pulping plants (chemical and mechanical) require proper water treatment to lower BOD and COD values before process- and wash-water is discharged into receiving waters. Any sulfides or sulfites in the effluent have to be oxidized to sulfate salts and the color has to be reduced to an acceptable level. Total suspended solids can be reduced in advance by coagulation, flocculation, sedimentation and if necessary, filtration.

Solid Waste

23. Preparing the wood for use in a pulping plant generates a great amount of solid waste: logging, including the felling of trees, clearing the tree of branches, and removal of bark, dirt, sand or stones. Proper disposal of this waste should be included in the design of the project. Other sources of solid waste are rejects from the screen, recausticizing rejects, wastewater sludges, and off-spec paper and trash. In addition, boiler ash may contribute as much as a fourth of the total solid waste products. Where possible, solid waste should be burned and the waste heat recovered. The burning of solid waste often has to be preceded by de-watering.

24. In practice, a pulpmill is often combined with a saw mill. The additional wood waste from a saw mill can be used in wood burning boilers or, in the case of sawdust and wood chips, as base material for the production of chipboard and wallboard.

25. Approximately 75 percent of the solid waste is organic and, if not burned, must be disposed of properly to avoid stress on the environment. Because toxic and hazardous solid wastes discharged into a landfill can degrade groundwater resources, proper disposal should be planned from the beginning. Lined storage with continuous leachate monitoring may be necessary. (For more information on "Solid Waste Collection and Disposal Systems," see Chapter 9.)

26. In developing countries, interest in the use of non-wood fibers in the pulp and paper industry is increasing. Small mills are being established using primarily rice and wheat straw, kenaf, bagasse, bamboo and jute cuttings, among other materials. From an environmental perspective, the most important difference between non-wood and wood raw materials is the high ash content of non-wood materials that create greater problems for solid waste disposal.

Waste Minimization

27. Pulp and paper production uses large quantities of water for washing and for pollution abatement equipment. To minimize the demand for outside water, a pulp and papermill should treat its wastewater and, if quality permits, recycle it in the process. To facilitate recycling, heavily contaminated streams in the process should be kept separate from lightly contaminated ones.

28. Solid waste disposal should be kept to a minimum. Solid wastes may be used as fuel in the plant for production of steam, though this may require cyclones and gas scrubbing equipment.

Project Alternatives

Site Selection

29. General issues for consideration in industrial plant siting are discussed in the section on "Plant Siting and Industrial Estate Development." Although the nature of pulp and paper production requires that special attention be given to the raw material supply, the mill should have a guaranteed wood supply near the plant site. Forests for harvesting wood should be identified and the environmental impacts considered in the design stage.

30. Another critical factor in site selection is the location of nearby towns and villages. For the prevailing wind directions, a pulp and paper plant should be situated downwind of towns and villages. Receiving waters with substandard quality already or with insufficient capacity to accept even well-treated effluents are inappropriate.

31. In some developing countries, agroindustry has fostered cooperation between farmers and the pulp and paper industry in the planting and maintenance of trees while growing crops on cleared patches of land. Some of these countries expect to obtain as much as 40 percent of their pulpwood each year from these "agro-forest" arrangements; site selection will be affected by such arrangements.

32. Finally, watershed management can be an important issue in the selection of a site. Bank policies on forest exploitation and watershed management are covered in the "Natural Forest Management" and "Watershed Development" sections of chapter 8.

Manufacturing Processes

33. A variety of alternatives for the production of paper pulp exist, but the possibilities are narrowed when certain types and qualities of paper have to be produced. Each process has been developed to meet specific performance and appearance criteria, as well as economic goals. Use of any of the processes will involve waste releases to the environment; however, the process will vary in terms of the quality and quantity of air pollutants, effluent water, and solid waste produced. During the design stage, end product criteria, technological constraints, and environmental objectives will define the viable alternatives to be considered. If, for example, only newspaper quality is required, then depending on the type of wood that is available, a mechanical pulping process may be sufficient with less impact on the environment. Another option may be the recycling of newspaper and other types of paper.

34. New processes to minimize wastes have been developed and some are already in operation. One is oxygen pulping, a process where no sulfur compounds are used and chlorination for bleaching the pulp can be eliminated. Although the paper quality is not yet equal to that of the kraft process, further research may eliminate this disadvantage. Another development is the Ranson process, a modified kraft process that operates as a closed system.

35. In the design phase, using by-products from other industries should be examined, for example, wood chips in chipboards, wood waste in wallboard, non-hazardous solid wastes in agricultural products, and so forth. In this connection, segregation of wastes at the source can be very efficient in waste reuse

systems. The following types of waste should be segregated: fibrous sludges, inorganic chemical sludges, bark, wood waste, ash, oil, hazardous chemicals, scrap metal, and biological sludges. Especially important is the separation of waste containing hazardous chemicals from bulky wastes.

Air Pollution Controls

36. Depending on the process and its location, one or more of the following may be required to bring air emissions to acceptable levels:

- electrostatic precipitators
- scrubbers
- cyclones
- wire mesh demisters
- filters
- incineration
- air or steam stripping
- liquid phase oxidation
- absorption

Water Quality Controls

37. Options for wastewater treatment and rinse include the following:

- water treatment and reuse
- dewatering of sludges
- evaporation
- sedimentation, flocculation, and filtration
- neutralization of acid or alkaline wastewater
- agricultural use
- dentrification

Management and Training

38. The potential negative impacts on water and air quality from the kraft, soda, and sulfite processes necessitate institutional support for effective pollution control management. The plant staff should include a plant engineer trained in pollution control of water and air and in monitoring technologies. Manufacturers will frequently supply the necessary equipment operations and maintenance training, if requested. Standard operation procedures and preventive maintenance should be established for the plant and enforced by management. They should include pollution control equipment operation and maintenance, air and water quality monitoring requirements, and instructions for notification and shutdown or other responses to pollution equipment failure.

39. Plant health and safety rules should be established and enforced. They should include:

- Provisions to prevent and respond to accidental release of hazardous chemicals (such as chlorine, ammonia, hydrogen sulfide), and spills of solutions and waste streams containing hazardous chemicals (sulfuric acid, sulfites, hypochlorites, peroxides).

- Procedures and monitoring to keep exposure to any of these chemicals in vapors and gases below World Bank limits.

- A program of routine medical examinations.

- Ongoing training in plant health and safety, and in good environmental housekeeping practices.

(For further discussion, see the section on "Industrial Hazard Management" and the World Bank's Occupational Health and Safety Guidelines.)

40. Emissions and effluent standards should be set for the plant based on national regulations where they exist, or on World Bank Guidelines where they do not. Government agencies charged with monitoring and enforcing standards should have the capability and authority to do so. The environmental assessment should include an evaluation of national and local capabilities and recommend appropriate elements of technical assistance to be included in the project.

Monitoring

41. Monitoring plans are necessary for plant/site specific. In general, however, a pulp and paper plant should include monitoring of:

- flue gas opacity
- emissions of particulates, chlorine, ammonia (if used), hydrogen sulfide, organic sulfur compounds (dimethyl sulfide, dimethyl disulfide) sulfur dioxide, nitrogen oxides
- workspace air quality for the same chemicals
- ambient air quality in vicinity of plants for applicable pollutants and odor
- process parameters that prove the operation of air pollution control equipment
- liquid waste streams for pH, TSS, sulfides, ammonia, sulfites, BOD_5 and COD,
- receiving water downstream and permitted stormwater discharges for dissolved oxygen, and applicable pollutants including particulates and pH
- working areas of all plants for ambient noise levels
- solid waste storage areas for runoff, infiltration and leachate (storage should be lined)
- inspection for safety and pollution control procedures

42. The forest management plan should specify monitoring of logging and harvesting practices to ensure compliance with environmental constraints. (For information on "Timber Harvesting," see Chapter 8.)

Table 10.17. Pulp, Paper, and Timber Processing

Potential Negative Impacts	Mitigating Measures

Direct: Site Selection

1. Siting of plant on/near sensitive habitats such as mangroves, estuaries, wetlands, coral reefs.

2. Siting along water courses causing their eventual degradation.

3. Siting can cause serious air pollution problems for local area.

Direct: Plant Operation

4. • Inadequate or non-existent forest management resulting in soil erosion, diminishing biotopes.

   • Unchecked pesticide application causing toxicologic effects on beneficial organisms and undesirable changes in forest ecosystems.

---

1. • Locate plant in industrially zoned area, if possible, to minimize or concentrate the stress on local environmental services and to facilitate the monitoring of discharges.

   • Integrate site selection process with natural resource agencies to review alternatives.

2. • Site selection process should examine alternatives that minimize environmental effects and not preclude beneficial use of the water body.

   • Plants with liquid discharges should only be located on a watercourse having adequate waste-absorbing capacity.

3. • Locate in an area not subject to air inversions or trapping of pollution, and where prevailing winds are towards relatively unpopulated areas.

4. • In project design phase, develop a forest management plan based on an environmental impact study.

   • Do not select wood supply from primary forest reserves (for further discussion, see sections on "Natural Forest Management" and "Tropical Forests").

Table 10.17. Pulp, Paper, and Timber Processing (continued)

Potential Negative Impacts	Mitigating Measures
Direct: Plant Operation (continued)	
5. • Release of gaseous wastes. • Sulfur dioxide • Total reduced sulfur compounds (TRS) • Particulates • Toxic organic compounds (e.g., chlorine, hydrogen sulfide)	5. • <u>Sulfur Dioxide</u> • Control by proper operations such as liquor recovery furnace. • Select appropriate auxiliary fuels. • Fuel desulfurization, flue gas scrubbing, and process modification. <u>TRS</u> • Collection by headers, scrubbed with alkali solution, then burned. <u>Particulate</u> • Removal by evaporator-scrubbers, cyclones or electrostatic precipitators. <u>Air toxins</u> • Prevent/control releases through process design.
6. • Release of liquid wastes to water bodies. • Conventional pollutants causing the following impacts: • changes in pH and toxicity • dissolved and suspended solids • eutrophication • foam and scum • slime growth • thermal effects • changes in taste, color and odor • fish-flesh tainting	6. • In-plant operating and housekeeping measures: • Pulp washing, chemical and fiber recovery, treatment and reuse of selected waste streams, collection of spills, and prevention of and collection tanks for accidental discharges. • Monitoring of sewers, drainage channels, and discharges to warn of spills. • Load leveling of treatment facilities by use of storage basins and other measures. • Recycling of barking water.

Table 10.17. Pulp, Paper, and Timber Processing (continued)

Potential Negative Impacts	Mitigating Measures
Direct: Plant Operation (continued)	

• Toxins such as trichlorophenol, pentachlorophenol and zinc.	• External effluent treatment:
	  • Primary—sedimentation basins, gravity clarifiers, and dissolved air flotation.
	  • Secondary—oxidation ponds, trickling filter, aerated lagoon, activated sludge, irrigation, sedimentation basin (to remove biological solids) and secondary clarifier.
	  • Toxins control by substitution of less/non-toxic chemicals.

7. • Disposal of solid wastes on the land.
  • Subsurface leaching with subsequent contamination of ground and surface water.
  • Destruction of ecologically sensitive areas such as marshes and other wetlands.
  • Proliferation of rodents, scavengers and insects harmful to human health.
  • Fires, health hazards, and unsightly conditions.
	7. • Source reduction, source segregation, by-product utilization, appropriate planning and management of disposal sites such as lining of disposal sites with collection system for run-off water and leachate (see "Solid Waste Collection and Disposal Systems" section).

8. • Sludge incineration.	8. • Dewatering by vacuum filtration and chemical conditioning to prepare sludges for burning.
	  • Incinerators:
	    • waste only
	    • burning in the bark boiler
	    • burning in power boiler

Table 10.17. Pulp, Paper, and Timber Processing (continued)

Potential Negative Impacts	Mitigating Measures
Indirect	
9. • Occupational health effects on workers due to: • Special pulp mill operations such as preparing logs (chipping and grinding). • Handling and storing of pulpwood and paper chips, and raw materials other than pulpwood. • Chemical processes used in making pulp, bleaching, and stock preparation. • Handling of spent liquors and machine room operations involves dust, fumes and gases, as well as special equipment such as shredders, clippers, cutters, heavy mobile equipment, etc.	9. • Facility should implement a Safety and Health Program designed to: • identify, evaluate, monitor, and control hazards to employees • design safe operating procedures • provide training in safety practices and the handling of emergencies
10. • Transit patterns disrupted, noise and congestion created, and pedestrian hazards aggravated by heavy trucks transporting raw materials, fuel and final products to/from the facility.	10. • Site selection can mitigate some of these problems. • Special transportation sector studies should be prepared during project feasibility stage to select best routes to reduce impacts. • Follow transportation regulations and develop emergency contingency plans to minimize risk of accidents.

MINING AND MINERAL PROCESSING

1. Projects in this sector involve mining, transportation, and processing of minerals and construction materials. These activities include:

- Surface and underground operations to produce metallic, nonmetallic, and industrial minerals, construction materials and fertilizers.

- *In situ* extraction of meltable and soluble minerals (notably sulfur, and, more recently, copper), dredging and hydraulic mining along rivers and coastal waters, and mine-site heap leaching (primarily gold and copper).

2. Material handling, within the mine area and to and from the processing facilities, require fleets of large excavating and transporting equipment (trucks, scrapers, shovels, draglines, bucket wheels and bulldozers), conveyors, pipelines, or rails. On-site processing facilities include preparation and washing plants for coal and construction materials, preparation plants, concentrators, leaching facilities at the mine-site and, depending on economics, on-site or off-site smelters and refineries. A large mining and/or milling operation is a major industrial complex, with up to thousands of workers, that requires an infrastructure consisting of utilities, air field, highways, railroad, port (if applicable), and all the usual ancillary community facilities.

Potential Environmental Impacts

3. All mining methods involve some disturbance of the surface and underlying strata, including aquifers. Exploration and pre-development impacts are usually short-term and include:

- surface disturbance from access roads, drill holes and test pits, and site preparation;
- airborne dust from road traffic, drilling, excavating, and site-clearing;
- noise and emissions from diesel equipment operation;
- disturbance of soil and vegetation, streams, drainages, wetlands, cultural or historic resources, and groundwater aquifers; and
- conflicts with other land uses.

4. Both surface and underground mining involve the following: drainage of the mine area and discharge of mine waters; the removal and storage/disposal of large volumes of waste material; and the removal and processing of the ore or construction material. Removal requires the use of diesel or electric powered mining and hauling equipment and a large skilled labor force. Extensive support facilities, such as a transportation complex, offices and equipment shops (some of this will be in the workings at underground mines), and utilities are needed. Ore transport within the mine area and to the processing facilities may be by truck, earth-hauler, rail, pipeline or conveyor belt, and will usually include bulk storage, blending, and loading facilities.

5. **Surface mines** include quarries, open pits, strip and contour mines, and mountain-top removal comprising a few hectares to several square kilometers. These operations require total disruption of the project area with large open pit(s) or quarry(s) and extensive overburden piles; however, it is often feasible to backfill the mined areas during or at the end of operations. Environmental concerns in surface mining include airborne particulates from road traffic, blasting, excavation and transport, emissions, noise, vibrations from diesel equipment and blasting, discharges of contaminated mine water, disruption of groundwater aquifers, removal of soil and vegetation, and visual intrusions. Other land uses at the site are precluded during the mining and reclamation activities. Slope or bench stability is a major concern in surface mining. Good mining practice requires constant observation for bench face movement that might indicate impending slope failure.

6. **Underground mining** methods include room-and-pillar, shrinkage stope, block caving, and longwall mining. These result in large voids under the land surface and piles of waste rock above ground; in many cases, however, portions of the underground workings are backfilled during mining. Most excavation is underground and requires the use of blasting equipment, but there will still be some surface operations. Possible impacts from underground mining include the removal of soil and vegetation, particulates, diesel emissions from surface operations, noise, vibrations from blasting, vented gases (blasting, diesel operations, radon), contaminated mine water discharge (nitrates, heavy metals, acidic, etc.), disruption of groundwater aquifers, fractures, ground instability or subsidence, and visual intrusions.

7. **Dredging and hydraulic mining** are usually conducted in alluvial material along the bed and banks of modern or ancient streams and in coastal areas or wetlands. Excavations and processing are done from either floating diesel-powered dredges (bucket-and-ladder, suction, or bucket-wheel) with pumps and first-stage processing facilities on board, shore-based diesel-powered draglines, conveyors, processing plant or hydraulic monitors (e.g., powerful jets of water that wash out the bank material), or with sluices to collect and direct the runoff, and separation facilities. These operations involve total disruption of the mined strata and modification of the local topography.

8. In dredging, mineral material is raised from the bottom by suction and/or mechanical excavators, and processed, rejects are discharged to the water or onshore. The bottom area is systematically swept during extraction with the dredge moving along the stream or coastline; stream channels are deepened and modified, wetlands and coastal areas are also deepened, resulting in large reject piles being left. In sand and gravel operations, the recovered material may be moved to shore by pipeline, conveyor or barge. Ores are usually concentrated onboard (mechanical, chemical, or amalgamation) and concentrate products or amalgam are shipped to shore for further upgrading and processing. Mercury, the amalgamation agent for gold and silver, presents particular environmental concerns and must be handled accordingly. In shore-based placer mining, there may be extensive mining of old river benches well above the level of the current stream bed.

9. *In situ* **leaching** requires an extensive surface network of close-spaced drill holes, pipelines and pumps to circulate a leachate (and after mineral extraction, a flushing and/or neutralizing solution) through the ore body. Operational problems include the loss of leachate control, piping, spills, leaks, incomplete flushing and/or neutralizing. Impacts include the disturbance of soil, vegetation, cultural and historic resources, air quality degradation from particulates and diesel emissions, contamination of groundwater by leachate, contamination of surface waters from spills, and noise from operations (drills,

traffic, pumps). *In situ* **leaching** requires a local transportation network, small skilled labor force, equipment (well drills, trucks, cranes, diesel generators, electric pumps), water, electric power supply, support facilities (office, shop, storage, and housing), airfield, and access roads.

10. **Heap leaching** may involve leaching of old dumps and workings as secondary recovery to an ongoing operation, or as is now common in low-grade disseminated gold deposits, leaching of newly mined material in large piles either on the surface or in old pits. Usually the land surface or pit bottom is prepared using liners and gravel; collection pipes are laid down and ore material piled over (the ore is usually from surface mining [see para 5]). Leachate (primarily sulfuric acid for copper and sodium cyanide for gold) is sprayed or ponded on the piles and collected for metal recovery. After leaching, the pile is flushed, percolated through to extract the metal, and/or neutralized before disposal.

11. Operational problems include leach pile stability, leachate control, wind and surface water erosion, leakage/seepage into surface water and groundwater, piping, and incomplete flushing, neutralizing and/or reclamation. In addition to the surface mining, impacts include air quality degradation by wind-blown particulates from leach piles, sedimentation of local drainages with leach pile material, surface water contamination by leaks and spills, groundwater contamination from liner leaks, loss of wildlife and livestock in leach ponds, and noise from pumps.

12. Processing facilities include preparation and wash plants, separation/concentration plants (gravity separation, leaching, amalgamation, ion exchange, floatation, etc.), refineries, and smelters. The ore processing facilities produce large amounts of wastes (tailings, slime, slag) to be disposed of on or near the site, and sometimes this can be returned to mined-out areas.

13. Environmental concerns include the disturbance of soil, vegetation, and local drainages during site preparation, air pollution from separation, concentrating and processing (fugitive dust and stack emissions), noise from transporting, transferring, crushing and grinding the ore, contamination of surface waters by spillage from mills or wash plants, contamination of local groundwater by leakage from tailings piles and slime ponds, contamination of local soils, vegetation, and surface waters by wind and water erosion of the waste piles, waste disposal, visual intrusion, and land-use conflicts.

14. Processing plants in mountainous districts often have difficulty finding areas to impound concentrator tailings and discharge these inert fines to fast-moving streams. Further downstream, they settle out in river bends, wide channels, floodplains, and shallow coastal waters. These fines adversely affect aquatic life and can cause damming and flooding of downstream communities. (For further discussion, see Table 10.18 at the end of this section.)

Natural Resource Issues

Water

15. Improperly cased or sealed drill holes may permit interchange and contamination between aquifers. Discharge from dewatering of surface or underground mines, without adequate neutralization or treatment, may be highly acidic and contaminate local surface waters and shallow groundwater with nitrates, heavy metals or oils from equipment, reduce local water supplies, or cause erosion of drainages

and stream channels. Removal of rock strata disrupts local aquifer continuity and can lead to interconnections and contamination between aquifers; backfill material can alter the hydraulic characteristics and water quality. Dredging and placer mining degrade surface water quality by greatly increasing suspended solids, reducing light transmission, and recirculating any contaminants in the bottom sediments. *In situ* mining may result in aquifer contamination through loss of control of leachate or failure to properly neutralize the leached zone at the end of operations.

16. Processing can degrade local surface waters by improper discharge of contaminated process waters by seepage or leaks from tailings ponds or pipelines, and from spills or improper disposal of solvents, lubricants, and process chemicals.

Air

17. Airborne particulates result from blasting, excavation and earth moving, transportation, material transfer, wind erosion of loose soil during surface mining, and any surface operations at underground mines. Nitrates from blasting and combustion products from operation of diesel equipment may be present at both surface and underground mines. Radon may be concentrated at vent stacks from underground mines. At dredging and *in situ* operations, combustion products from diesel equipment will be present. During processing, airborne particulates result from transport, reduction (screening, crushing, or pulverizing), equipment traffic, wind erosion from dry areas of the tailings pond, roads, and material stockpiles.

Land

18. In surface mining the excavation, overburden disposal or storage, and construction of ancillary facilities results in removal or covering of soils and vegetation, disruption or blockage of streams, drainages, wetlands or coastal area, and extensive modification of the topography over the entire mined area. In dredging and placer mining, these effects are concentrated on water areas: stream channels may be rerouted, residual ponds created, and beaches eliminated; adjoining stream banks or shorelines may be used for waste disposal and ancillary facilities.

19. Underground mining requires land for waste rock disposal, storage of ore and low-grade material, and siting of ancillary surface facilities with effects similar to those listed above for surface mining. Land over the workings will be unstable, with fracturing and subsidence. Mining may result in the loss or modification of soils, vegetation, wildlife habitat, drainages, wetlands, cultural and historic resources, survey markers, topographic features, temporary or permanent loss of land productivity, and contamination of soils from mineral materials and toxic substances.

Sociocultural Issues

Land Use

20. Mineral exploration is a short-term and locally intensive land use that will conflict to some degree with existing non-mineral land uses. These activities can be serviced by air in remote areas, thus elimi-

nating the need for and intrusion of access roads. Surface mines, processing facilities, heap leach operations, *in situ* operations, and surface operations at underground mines require total occupation and use of the sites, precluding other uses. At underground mines, the surface used over the workings will be constrained by the degree of subsidence risk (which can range from negligible to certain), and depending on geology, ore depth, and mining method. Post-mining land use suitability will depend on the type, degree, and success of reclamation.

21. New access roads, utilities, and townsite(s) not only encourages additional population influx and secondary development peripheral to or unrelated to the mineral activities, but they can also initiate unplanned development and modification or degradation of remote areas that may continue after the mineral project ceases.

Cultural Resources

22. Surface disturbance from mining operations and related construction may damage or destroy cultural resources, historic sites, or native religious sites. Increased human presence in the area may lead to vandalization of unprotected sites.

People

23. Exploration and mining activities will interfere to some degree with other activities that may be present or planned in the area; for example, vibrations from equipment operation and blasting, as well as noise and dust can be distracting and may cause health problems to workers and nearby residents. The immigration of workers and their families may strain community services and cause "boom and bust" economic, social, or cultural conflicts, or even displace local populations. The initial construction force is usually transient and soon replaced by a smaller, permanent operations staff. (See the section on "Induced Development" in Chapter 3.)

Special Issues

Reclamation

24. Contemporaneous or post-mining reclamation of the site for other uses may not be feasible. Residual problems from surface mining may include erosion, weathering, water-logging, as well as the failure of remaining steep highwalls and waste pile slopes, and safety hazards from water-filled pits.

25. Residual problems from underground mining may include slumpage and the collapse of poorly supported workings that can lead to surface fractures, voids, and collapses; abandoned operations can create an attractive hazard, especially to children.

26. Other problems associated with surface and underground mining include the following:

- Fire in improperly sealed or reclaimed coal seams. This is common where remnant open pit coal mine is used for trash disposal (and burning), and can lead to CO emission, fracture, and collapse of the land surface.

- Groundwater seepage from the abandoned mine workings that can be highly acidic and/or contaminated with hazardous metals.

- Disruption of aquifers through actual removal or from mining-induced fracturing that can result in loss or degradation of local groundwater supplies.

- Down-slope damages from failure of waste rock piles located on steep slopes.

- Residual hazardous mineral materials exposed in surface workings and scattered in waste dumps.

In Situ and Heap Leaching

27. Major concerns at *in situ* leaching operations include groundwater contamination from loss of control (or excursions) of solutions injected and recovered, or from failure to adequately neutralize the leached zone or pile after operations. Additional concerns at above-ground heap leach operations include pile stability, access to leach ponds (by livestock and wildlife, especially birds), and fugitive dust from dried portions of the leach pile.

28. At processing sites, the major concerns center on the mill tailings pile or pond are: (a) seepage of highly contaminated water from the pile may contaminate surface and/or groundwater; (b) eroding and slumping of the sides of the pile can result in contaminating local soils and vegetation; (c) pH and/or residual mineral content may hamper revegetation; (d) particulates from wind erosion may pose air pollution hazard; and (e) residual slime may take years to dry.

Dredging

29. In dredging and large-scale placer mining, permanent damage may occur to fisheries, water quality, and aesthetics from modifying the course and flow characteristics of a natural free-flowing stream, in addition to lining the banks with rock piles, and inundating downstream areas with sediments. During operations, other water users may be restricted and quality of water to downstream users may be greatly degraded.

Project Alternatives

30. Other than the alternative of "no action" or not going forward with all or parts of the project, alternatives for the mining are generally determined by the type and degree of mitigation that will be required. Mitigation may be specially tailored to the particular operation. The mining method (surface, underground, *in situ*, or dredge) is determined for the most part by economics, depth, configuration, grade, and mineral characteristics of the ore body, and the geology of the host rock.

31. There may be some flexibility in the placement of the waste piles, post-mining reclamation, mining equipment, and transport of ore from mine to mill. There is usually considerable leeway as to mill location and placement of tailings piles and/or ponds, although from an economic standpoint, it is generally desirable to minimize ore transport distance from mine to mill and tailings pipeline from mill to pond. Critical factors in mill location include adequate water supply, site for tailings pond(s), and transportation access. Processing methods are determined by the mineral and host rock characteristics, economics, and availability of water. A small mine may ship to a custom mill, if available; large mines may include a refinery (secondary processing) with the concentrator.

32. In general, mitigation measures may include avoidance of areas with sensitive resources, restriction on timing of operations, siting of structures, or utility/transportation corridors to avoid resource conflicts, controlled rate of development, or phased development to limit socioeconomic impacts, and special resource or community studies as basis for subsequent mitigation. Proper engineering in the design of roads, surface excavations, piles (waste, tailings, leach), surface ponds, mine drainage, underground workings, and structures can enhance safety and reduce incidence of accidents.

Management and Training

33. Basic to safe mining operations, i.e., protection of workers, the general public and the environment, are adequate regulations with a competent inspection staff and effective regulations and enforcement. Adequate training in equipment operation and ongoing intensive safety programs are essential to minimizing accidental injuries and/or fatalities. Training and safety requirements for the mining staff are similar to those of the construction industry, and the requirements for mill workers are similar to those of an industrial chemical plant. (See the "Industrial Hazard Management" section for further discussion.)

Monitoring

34. At surface mines, standards may be set and monitoring may be required for air quality (particulates and equipment emissions), groundwater (drawdown), seismic (blast) vibrations, pit wall slope and stability, surface water flow and quality (especially sediment), mine drainage, sanitary wastes, identification and separate disposal of non-economic mineralized material encountered in mining, management and disposal of hazardous wastes, and radioactivity levels at workings and project boundaries.

35. Reclamation requirements and monitoring include restoration of land surface (drainage, slope, stability), revegetation (cover, type, vigor), groundwater (recovery, quality), surface water quality, and surface radon emissions. At underground mines, much of the above will apply with some additions, such as earth movement (especially at faults, major fractures, and subsidence zones), and air quality at vents.

36. At dredging and placer mining, operations standards and monitoring will include equipment emissions, sediment control, discharge water quality, amount and timing of stream diversion, and waste material discharge (method, location, configuration).

37. At *in situ* operations, monitoring requirements include perimeter monitoring wells (especially downgradient), input/output of leachate, well tests (pressure, proper sealing), and visual checks for leachate spills or leaks (pipelines, transfer points, and storage tanks).

38. Reclamation monitoring includes an analysis of return flushing or neutralizing solutions, verifying proper removal and/or sealing of wells, and periodic testing of perimeter monitoring wells. Above-ground heap leach operations have additional requirements, such as visual checking of pile slopes and toe for leakage, downgradient surface water sampling, and reclamation of pile after flushing and/or neutralizing (slope and vegetative cover).

39. At processing operations, monitoring requirements should include the following: air quality at stacks, on-site and facility boundaries, quantity and quality of water discharge, identification and proper handling of hazardous wastes, and noise levels both on-site and at facility boundaries.

Table 10.18. Mining and Mineral Processes

Potential Negative Impacts	Mitigating Measures
Direct	

1. • Modification/loss of soil profile, vegetation, and surface drainages during exploration, mining, and construction.	1. • Require appropriate resource surveys, before disturbance, of areas that may be affected by the project to identify:
	    • cultural and historic resources
	    • flora and fauna
	    • soils
	    • surface and groundwater quality and quantity
	    • land uses
	    • significant topographic features
	
	• Mitigation measures based on identified resource conflicts may include:
	    • avoidance
	    • timing of operations
	    • recovering and archiving cultural and historic resources
	    • segregation and stockpiling for use in reclamation (soils)

2. • Damage/destruction of cultural resources and historic sites. • Survey monuments during exploration, mining and construction.	2. See No. 1.
3. • Degradation of surface waters by soil erosion from disturbed areas, waste piles, and stockpiles. • Decreased capacity of local reservoirs/ponds from siltation.	3. • Require control of stormwater runoff and prompt revegetation on disturbed areas. • Avoid disturbance of streams, drainages, ponds and wetlands.

Table 10.18. Mining and Mineral Processes (continued)

Potential Negative Impacts	Mitigating Measures
Direct (continued)	
	• Where disturbance cannot be avoided, require use of sediment control structures/practices.
	• Water quality standards should include suspended solids.

4. Contamination of surface waters and shallow groundwater wastes, (aquifers) by waste water from mine drainage, equipment servicing, and sanitary and domestic wastes.	4. • Require treatment of:
	   • mine drainage
	   • sanitary/domestic and stormwater runoff to meet water quality standards before discharge
	 • Prompt cleanup of any spills (oils, lubricants and cleaning solvents).
	 • Water quality standards should be established for all waste water discharges.

5. Disruption/contamination of local aquifers by exploration drill holes and mining excavations.	5. • Avoid or minimize penetration of aquifers below the strata being mined. • Drill holes outside or below the mine area should be properly cased or sealed.
6. Reduction in local water supplies.	6. Require replacement from alternate sources.

Table 10.18. Mining and Mineral Processes (continued)

Potential Negative Impacts	Mitigating Measures

Direct (continued)

7. Reduced reproduction and populations of wildlife from habitat modification and loss.

    7. • Prohibit or restrict disturbance of significant habitat wetlands.
        • Require prompt reclamation to forage and habitat favorable to local wildlife.

8. Wildlife mortality from road traffic and surface disturbances.

    8. • Mark wildlife road crossings.
        • Emphasize driver awareness.
        • Install road underpasses.

9. Degradation/loss of vegetation (and soil productivity) from discharge of contaminated waters (see No. 4).

    9. See No. 4.

10. Modification of vegetation and introduction of non-native species.

    10. Require prompt reclamation of disturbed areas and revegetation with native species.

11. Contamination of surface areas with mineralized/toxic rock material.

    11. Require identification and segregation of toxic rock materials.

12. Degradation of air quality and visibility from airborne particulates (blasting, road traffic, wind erosion).

    12. • Require the following:
           • proper blasting practices to minimize airborne particulates
           • watering haulage roads
           • prompt revegetation or application of sealants and dust suppressants to disturbed areas (including waste and topsoil piles)

Table 10.18. Mining and Mineral Processes (continued)

Potential Negative Impacts	Mitigating Measures
Direct (continued)	
13. Degradation of air quality from routine operational (diesel) emissions.	13. • Appropriate pollution control devices should be installed and operative on all diesel/gasoline powered equipment. • Hydrocarbon vapor control at all fuel transfer points. • Prompt cleanup of any oil spills.
14. Air quality degradation from processing emissions.	14. • Require use of adequate technology to ensure emissions are kept at acceptable levels.
15. Land-use conflicts.	15. • Consult with local land users in siting access roads, air fields, utility lines, and to extent possible, mining and processing facilities. • Allow other land uses to continue on the site where compatible with the operations.
16. Road damage, accidents, and traffic delays from increased truck traffic on local roads.	16. • Observe road load limits. • Design roads for adequate capacity and visibility. • Ensure that roads are properly signed, vehicles are well-maintained, and drivers are trained and safety-conscious. • Provide buses or require that commuting workers car-pool or provide buses.

Table 10.18. Mining and Mineral Processes (continued)

Potential Negative Impacts	Mitigating Measures
Direct (continued)	
17. Visual intrusions from drill rigs, surface mine excavations and equipment, mine facilities and headframes (underground mines). Cleared linear rights-of-way for pipelines, utilities, roads and processing facilities (see also No. 12).	17. • Paint structures to blend with background (vegetation and sky). • Avoid contrasting colors. • Utilize utility corridors, minimize clearing, and blend vegetation where feasible.
18. Disturbance of humans and wildlife by noise from equipment operation, blasting, and processing facilities.	18. • Utilize earth mound vegetative screening. • Follow proper blasting procedures, use minimum charges, and avoid blasting at night or early morning.
19. Damage to structures and disturbance of local residents by blasting vibrations.	19. • Use blasting procedures to minimize vibrations to nearby residences and structures, and install monitoring instruments at sensitive locations.
20. Injury/loss of life from accidents.	20. • Periodic training and continual safety reminders to all operating staff. • Require periodic drills in emergency procedures. • Ensure that all visitors are briefed on potential hazards and necessary safety precautions. • Ensure that appropriate safety and rescue equipment is available and employees trained in its use.

Table 10.18. Mining and Mineral Processes (continued)

Potential Negative Impacts	Mitigating Measures

Direct (continued)

21. Increased demands on services and facilities in local communities, social and cultural conflicts, concern with community stability (boom and bust scenario).

21.
- Require pre-development, socioeconomic study of potentially affected communities to identify possible impacts on services, infrastructure, dislocations, and conflicts.
- These impacts can be addressed by:
  - community assistance grants
  - loans
  - prepayment of taxes
  - phasing mineral development
  - constructing needed community facilities
- Cooperative and open working relations should be established early with local communities and maintained throughout the life of the project.
- Project workers should be encouraged to participate in community affairs.

22. Conflicts with native cultures, traditions, and life-styles.

22.
- Brief all employees to ensure awareness of and sensitivity to the local cultures, traditions, and lifestyles.
- Ensure that native leaders are aware of the projected activities, are assisted in identifying impacts that may be of particular concern to them, and have a voice in appropriate mitigation measures.
- Mitigation may include isolating the work force from the native community.

Table 10.18. Mining and Mineral Processes (continued)

Potential Negative Impacts	Mitigating Measures
Direct (continued)	
23. Subsidence of land surface (underground mining).	• Require adequate support be provided in the underground workings through pillars, cribbing or backfill.
	• Monitor controlled subsidence and identify possible subsidence areas for land-use restrictions.
24. Loss of birds and animals in tailings and leach ponds.	• Minimize surface area of tailings and leach ponds, and require that they be promptly drained or closed when not in use.
	• Net covering, fencing, or scaring may be required at active ponds.
25. Modification/disruption of surface waters (dredging).	• Require use of sediment control structures/practices.
	• Water quality standards should include suspended solids.
Indirect	
1. Degradation of remote areas through improved access and increased use.	• Access remote areas by air rather than roads during early exploration stage.
	• Restrict use of access roads, and remove and reclaim any access roads at end of production.
	• Minimize need for community development by rotating work crews and precluding permanent residences.

Table 10.18. Mining and Mineral Processes (continued)

Potential Negative Impacts	Mitigating Measures
Indirect (continued)	
2. Vandalization of cultural resources and historic sites.	2. • Do not publicize cultural resource sites in remote or unprotected locations. • Restrict unnecessary access and patrol sites.
3. Wildlife loss through poaching.	3. Prohibit carrying of firearms in area, restrict unnecessary access, and patrol areas.
4. Secondary population growth and related effects.	4. See No. 20.

References

Industrial Hazard Management

American Conference of Governmental Industrial Hygienists. 1977. Threshold Limit Values and Biological Exposure Indices. Updated annually. Cincinnati, Ohio.

Asian Development Bank. 1988. Environmental Guidelines for Selected Industrial and Power Development Projects. Manila, Philippines: Environment Unit, Infrastructure Department.

Batstone, R., J. E. Smith, Jr., and D. Wilson, eds. 1989. The Safe Disposal of Hazardous Wastes: The Special Needs and Problems of Developing Countries. World Bank Technical Paper 93. Washington, D.C.: World Bank.

United Nations Environment Programme. 1988. Awareness and Preparedness for Emergencies at Local Level (APELL): A Process for Responding to Technological Accidents. Nairobi, Kenya.

World Bank. 1978. Environmental Considerations for the Industrial Development Sector. Office of Environmental and Health Affairs. Washington, D.C.: World Bank.

___. 1988. Environmental Guidelines. Environment Department. Washington, D.C.: World Bank.

___. 1988. Occupational Health and Safety Guidelines. Environment Department. Washington, D.C.: World Bank.

___. 1988. Techniques for Assessing Industrial Hazards. Technical Paper 55. Washington, D.C.: World Bank.

Hazardous Materials Management

Anderson, I. "White Asbestos Also Causes Canacer," New Scientist, March 9, 1991, p. 12.

Huncharek, M. 1990. "Brake Mechanics, Asbestos and Disease Risk." American Journal of Forensic Medical Pathology 11(3):240-326.

Mossman, B. T., and J. B. Gee. 1989. "Asbestos-Related Diseases." New England Journal of Medicine 320:1721-1730.

Mossman, B. T. and others. 1990. "Asbestos: Scientific Developments and Implications for Public Policy." Science 247:361-365.

National Institute for Occupational Safety and Health. 1985. <u>Occupational Safety and Health Guidance Manual for Hazardous Waste Site Activities</u>. Washington, D.C.: United States Department of Health and Human Services. <u>Bibliography</u>. Document No. PB 90-87555. Washington, D.C.

Neuberger, M., and M. Kundi. 1990. "Individual Asbestos Exposure. Smoking and Mortality: Asbestos Cement Industry." <u>British Journal Of Industrial Medicine</u> 47:615-620.

Occupational Safety and Health Administration. 1990. "Occupational Safety and Health Standards, Subpart Z. Toxic and Hazardous Substances: Asbestos, Tremolite, Anthrophyllite and Actinolite." <u>Federal Register</u> 1910.1001. <u>U.S. Standards and Interpretations (OSHA)</u> 702.0.32-708.0.29. Washington, D.C.: United States Department of Health and Human Services.

Pelnar, P. V. 1990. <u>Health Effects of Asbestos and of Other Minerals and Fibers</u>. Park Forest, Illinois: Chem-Orbital.

United States Printing Office. 1990. "Hazardous Waste Operations and Emergency Response." <u>U.S. Standards and Interpretations</u>. <u>Federal Register</u> 1910.132. Washington, D.C.

Plant Siting and Industrial Estate Development

Geethkrishnan, K. P. 1989. "Indian Policy for Siting of Industry." <u>Industry and Environment</u> 12(2)3-6.

United Nations Environment Programme. 1980. <u>Guidelines for Assessing Industrial Impact and Environmental Criteria for the Siting of Industry</u>. 2 Volumes. Paris, France: Industry and Environment Office.

United States Environment Programme. 1980. "Guidelines to Assessing Industrial Environmental Impact and Environmental Criteria for the Siting of Industry." <u>Industry and Environment Guidelines Series</u>, Volume 1. Paris, France.

Electric Power Transmission Systems

Asplundh Environmental Services. 1979. <u>Right-of-Way Ecological Effects Bibliography</u>. Report No. EPRI-EA-1080. Willow Grove, Pennsylvania.

Gas Research Institute. 1988. <u>Environmental Aspects of Rights-of-Way for Natural Gas Transmission Pipelines: An Updated Bibliography</u>. Argonne, Illinois: Energy and Environmental Systems Division.

Goodland, R., ed. 1973. <u>Power Lines and the Environment</u>. Millbrook, New York: Cary Ecosystem Center.

United States Department of the Interior. 1979. <u>Environmental Criteria for Electric Transmission Systems</u>. Document No. 001-010-00074-3. Washington, D.C.: General Printing Office.

United States Environmental Protection Agency. 1980. <u>Electric Fields Under Power Lines</u>. Supplement to an Examination of Electric Fields Under EHV Overhead Power Tansmission Lines. Silver Spring, Maryland.

United States Fish and Wildlife Service. 1979. <u>Management of Transmission Line Rights-of-Way for Fish and Wildlife. Volume I: Background Information</u>. Report No. FWS/OBS-79/22-1.

Oil and Gas Pipelines

Carpenter, G. F. 1982. "Environmental Considerations in Planning and Routing Natural Gas Pipelines and Their Relationship to Off-Road Vehicle Use." <u>Right of Way</u> 29(3):29-31.

Considine, C. M., ed. 1977. "Gas Pipelines and Underground Storage," in <u>Energy Technology Handbook</u>. New York: McGraw-Hill.

Davis, S. H., Jr. 1976. "Effect of Natural Gas on Trees and Other Vegetation." <u>Journal of Aboriculture</u> 3(8):153-168.

Gas Research Institute. 1982. <u>Assessment of Environmental Problems Associated With Installation and Maintenance of New Gas-Transmission Pipelines</u>. Report No. GRI-81/0114. Chicago, Illinois.

Gosselink, J. G. 1984. <u>Impacts of Pipeline Installation in Coastal Louisiana</u>. Minerals Management Service Information Transfer Meeting. Metairie, Louisiana.

___. 1981. <u>Managing Oil and Gas Activities in Coastal Environments: Refuge Manual</u>. Report No. FWS/OBS-81/22. Austin, Texas.

World Bank. 1988. <u>Environmental Guidelines</u>. Environment Department. Washington, D.C.: World Bank.

Oil and Gas Pipelines--Offshore

United States Environmental Protection Agency. 1980. <u>Choosing Offshore Pipeline Routes: Problems and Solutions</u>. Prepared by the New England River Basins Commission. Report No. EPA-600/7-80-114. Boston, Massachusetts.

United States Fish and Wildlife Service. 1979. <u>Environmental Planning for Offshore Oil and Gas. Volume I: Recovery Technology</u>. Prepared by the Conservation Foundation. Report No. FWS/OBS-77/12. Washington, D.C.

___. 1978. Environmental Planning for Offshore Oil and Gas. Volume II: Effects on Coastal Communities. Prepared by the Conservation Foundation. Report No. FWS/OBS-77/13. Washington, D.C.

___. 1978. Environmental Planning for Offshore Oil and Gas. Volume III: Effects on Living Resources and Habitats. Prepared by the Conservation Foundation. Report No. FWS/OBS-77/14. Washington, D.C.

Oil and Gas Development--Onshore

Naval Civil Engineering Laboratory. 1981. Nearshore Pipeline Installation Methods. Report No. CEL-CR-81.016. Port Hueneme, California.

United States Department of Agriculture. 1982. Final Environmental Impact Statement for Proposed Oil and Gas Drilling at Cache Creek-Bear Thrust, near Jackson, Teton County, Wyoming. Washington, D.C.: U.S. Department of the Interior and Forest Service.

United States Department of the Interior. 1987. Hickey Mountain-Table Mountain Oil and Gas Field Development: Final Environmental Impact Statement and Record of Decision. Washington, D.C.: Bureau of Land Management, United States Department of the Interior.

___. 1981. Oil and Gas Environment Assessment of BLM Leasing Program: Lewistown District. Washington, D.C.: Bureau of Land Management, United States Department of the Interior.

Hydroelectric Projects

Dixon, J. A., L. M. Talbot, and G. J-M. Le Moigne. 1989. Dams and the Environment: Considerations in World Bank Projects. World Bank Technical Paper 110. Washington, D.C.: World Bank.

Garzon, C. 1984. Water Quality in Hydroelectric Projects: Considerations for Planning in Tropical Forest Regions. World Bank Technical Paper 20. Washington, D.C.: World Bank.

Goodland, R. 1978. Environmental Assessment of the Tucurui Hydroproject, Amazonia. Brasilia, Brazil: Eletronorte.

___. 1989. "The World Bank's New Policy on the Environmental Aspects of Dam and Reservoir Projects." Indian Journal of Public Administration 35(3):607-633.

Intermin Mekong Committee. 1982. Environmental Impact Assessment: Guidelines for Application to Tropical River Basin Development. United Nations Economic and Social Commission for Asia and the Pacific Environment and Development Series. Bangkok, Thailand.

Mahmood, K. 1987. Reservoir Sedimentation: Impact, Extent, and Mitigation. World Bank Technical Paper 71. Washington, D.C.: World Bank.

World Bank. 1989. "Environmental Policy for Dam and Reservoir Projects." Operational Directive 4.00, Annex B. World Bank, Washington, D.C.

Thermoelectric Projects

United States Fish and Wildlife Service. 1978. A Biologist's Manual for the Evaluation of Impacts of Coal-Fired Power Plants on Fish, Wildlife, and Their Habitats. Report No. FWS/OBS-78/75. Argonne, Illinois: National Laboratory.

___. 1979. Impacts of Coal-Fired Power Plants on Fish, Wildlife, and Their Habitats. Report No. FWS/OBS-78/29. Argonne, Illinois: National Laboratory, Division of Environmental Impact Studies.

Financing Nuclear Power: Options for the Bank

Ahearne, J. F. 1989. "Will Nuclear Power Recover in a Greenhouse?" Resources for the Future, ENR 89-06:58.

Blix, H. 1990. "The World's Energy Needs and the Nuclear Power Option." IAEA Bulletin 32(1):38-44.

Criqui, P. 1989. "Trends in World Energy Demand in the Face of Possible Global Climate Change." Energy Studies Review 1(3):258-268.

Cruver, P. C. 1989. "How the 'Greenhouse Effect' Might Shape World Energy Policy in the 21st Century." OPEC Bulletin 20(4):9-13.

De, P. L. 1990. "Costs of Decommissioning Nuclear Power Plants." IAIA Bulletin 32(3):39-42.

McKee, K. C. 1990. "The Lessons of Three Mile Island." Public Utilities Fortnightly 126(11):15-21.

Miller, A., and I. Mintzer. 1990. "Global Warming: No Nuclear Fix." Bulletin Atomic Scientists 46:30-34.

Murray, J. 1990. "Can Nuclear Energy Slow Global Warming?" Energy Policy 18:494-499.

Todani, K., Y. M. Park, and G. H. Stevens. 1989. "The Near Term Contribution of Nuclear Energy in Reducing CO_2 Emissions in OECD Countries." Energies Technologies for Reducing Emissions of Greenhouse Gases 1:59-70.

Cement

Beers, A. 1987. <u>Hazardous Waste Incineration: The Cement Kiln Option</u>. New York: State Legislative Commission on Toxic Substances and Hazardous Wastes.

Occupational Safety and Health Association. 1984. <u>Industrial Hygiene Technical Manual. Occupational Safety and Health Administration Instructions</u>. Washington, D.C.: General Printing Office.

United Nations Industrial Development Organization. 1977. <u>Information Sources on the Cement and Concrete Industry</u>. Guides to Information Sources, No. 2. New York.

United States Environmental Protection Agency. 1974. <u>Development Document for Effluent Limitations, Guidelines, New Source Performance Standards for the Cement Manufacturing Point Source Category</u>. Document No. EPA/440/1-74-005a. Washington, D.C.

World Bank. 1984. <u>Cement Manufacturing: Guidelines for Disposal of Waste</u>. Office of Environmental Affairs. Washington, D.C.: World Bank.

Chemical and Petrochemical

American Petroleum Institute. 1973. <u>Guidelines on Noise: A Medical Report</u>. Austin, Texas.

National Institute for Occupational Safety and Health. 1985. <u>Guide to Chemical Hazards</u>. No. 78-210. Washington, D.C.: United States Department of Health and Human Services.

United States Environmental Protection Agency. <u>Effluent Guidelines and Standards for Organic Chemicals</u>, (40 CFR 414).

___. <u>Effluent Guidelines and Standards for Inorganic Chemicals</u>, (40 CFR 415).

___. <u>Effluent Guidelines and Standards for Pharmaceutical Manufacturing</u>, (40 CFR 439).

___. <u>Regulations on National Emission Standards for Hazardous Air Pollutants</u>, (40 CFR 61).

Fertilizer

United Nations Environment Programme. 1982. <u>Guidelines for Assessing Industrial Environmental Impact and Environmental Criterial for the Siting of Industry</u>. 2 Volumes. Paris, France: Industry and Environmental Office.

United States Environmental Protection Agency. <u>EPA Effluent Guidelines and Standards for Fertilizers Manufacturing</u>, (40 CFR 418).

World Bank. 1983. <u>Guidelines for Fertilizer Manufacturing Wastes</u>. Office of Environmental Affairs. Washington, D.C.: World Bank.

___. 1988. <u>Occupational Health and Safety Guidelines</u>. Washington, D.C.: World Bank.

Food Processing

United States Environmental Protection Agency. <u>Effluent Guidelines and Standards for Canned and Preserved Fruits and Vegetables</u>, (40 CFR 407).

___. <u>Effluent Guidelines for Canned and Preserved Seafood</u>, (40 CFR 408).

___. <u>Effluent Guidelines and Standards for Diary Products</u>, (40 CFR 409).

___. <u>Effluent Guidelines and Standards for Grain Mills</u>, (40 CFR 406).

___. <u>Effluent Guidelines and Standards for Meat Products</u>, (40 CFR 432).

___. <u>Effluent Guidelines for Sugar Processing</u>, (40 CFR 465).

World Bank. 1983. <u>Fruit and Vegetable Processing Industrial Waste Disposal</u>. Office of Environmental Affairs. Washington, D.C.: World Bank.

___. 1983. <u>Meat Processing and Rendering Industrial Waste Disposal</u>. Office of Environmental Affairs. Washington, D.C.: World Bank.

Iron and Steel Manufacturing

Kendrick, D. A., A. Meeraus, and J. Alatore. 1984. <u>Planning of Investment Programs in the Steel Industry</u>. Volume III. Baltimore, Maryland: The John Hopkins University Press.

Schueneman, Jean J., M. D. High, and W. E. Bye. 1963. <u>Air Pollution Aspects of the Iron and Steel Industry</u>. Public Health Series Publication 999-AP-1. Washington, D.C.: United States Department of Health and Human Services.

United States Environmental Protection Agency. <u>Effluent Guidelines for Iron and Steel Manufacturing</u>, (40 CFR 420).

World Bank. 1983. <u>Effluent Guidelines for Iron and Steel Industry</u>. Office of Environmental Affairs. Washington, D.C.: World Bank.

Nonferrous Metals

United States Environmental Protection Agency. Effluent Guidelines and Standards for Non-ferrous Metals, (40 CFR 421).

___. Regulations on National Emission Standards for Hazardous Air Pollutants, (40 CFR 61).

___. Regulations on Standards of Performance for New Stationary Sources, (40 CFR 60).

Petroleum Refining

Scherr, R. 1991. "Impact of Clean Air Act Amendment on Refinery Planning and Construction." Paper presented at the 89th Annual Meeting of the National Petroleum Refiners Association. San Antonio, Texas.

United States Environmental Protection Agency. Effluent Guidelines and Standards for Petroleum Refining, (40 CFR 491).

World Bank. 1988. Environmental Guidelines. Environment Department. Washington, D.C.: World Bank.

Pulp, Paper, and Timber Processing

Htun, N. 1982. International Trends in Environmental Management in the Pulp and Paper Industry. Bangkok, Thailand: Technical Association of the Pulp and Paper Industry Press.

Jensen, W. 1986. "Environmental Trends in the Finnish Pulp and Paper Industry." Industry and Environment 9(3):19-24.

United Nations Environment Programme. 1982. Environmental Guidelines for Pulp and Paper Industry. Environmental Management Guidelines No. 4. Nairobi, Kenya.

United States Environmental Protection Agency. 1980. Development Document for Effluent Limitations Guidelines and Standards for the Pulp, Paper, and Paperboard and the Builder's Paper and Board Mills. Report No. EPA 440/1-80/025b. Washington, D.C.: General Printing Office.

World Bank. 1980. Environmental Considerations in the Pulp and Paper Industry. Office of Environmental Affairs. Washington, D.C.: World Bank.

Mining and Mineral Processing

Cardamone, M. A., J. R. Taylor, and W. J. Mitsch. 1984. <u>Wetlands and Coal Surface Mining: A Management Handbook</u>. Kentucky: University of Louisville, Systems Science Institute.

Chrinonis, N. P., ed. 1980. <u>Training Manual for Miners</u>. New York: McGraw-Hill.

Hinkle, C. R., R. F. Ambrose, and C. R. Wenzel. 1981. <u>A Handbook for Meeting Fish and Wildlife Information Needs to Surface Mine Coal: OSM Region I</u>. Prepared for the Office of Surface Mining Reclamation and Enforcement and Office of Biological Services, Fish and Wildlife Service. Washington, D.C.: United States Department of the Interior.

Lamb, A. M. 1982. <u>Procedures for Assessment of Cummulative Impacts of Surface Mining of the Hydrologic Balance</u>. Washington, D.C.: Office of Surface Mining Reclamation and Enforcement, United States Department of the Interior.

Lyle, F. S. 1987. <u>Surface Mine Reclamation Manual</u>. Washington, D.C.: Bureau of Mines, United States Department of the Interior.

Richings, M. L., and L. Readdy. 1981. <u>Surface Oil Mining: A Technical and Environmental Assessment</u>. Washington, D.C.: Bureau of Mines, United States Department of the Interior.

ANNEX 10-1

Sample Terms of Reference (TOR)
An Environmental Assessment of Energy Facilities

Note: Paragraph numbers correspond to those in the
Sample Terms of Reference (TOR) Outline in
Annex 1-3; additional paragraphs are not numbered

1. <u>Introduction</u>. This section will state the purpose of the terms of reference, identify the energy development projects to be assessed, and explain the executing arrangements for the environmental assessment. Energy development projects include, but are not limited to: electric power transmission systems, oil and gas pipelines, oil and gas development, geothermal development, hydroelectric facilities, and thermoelectric power plants.

2. <u>Background Information</u>. This section will provide pertinent background for potential parties who may conduct the environmental assessment, whether they are consultants or government agencies. The section will include a brief description of the major components of the proposed project, a statement of the need for it and the objectives it is intended to meet, the implementing agency, a brief history of the project (including alternatives considered), its current status and timetable, and the identities of any associated projects. If there are other projects in progress or planned within the region which may compete for the same resources, identify them within this section.

   Major components of an energy project to be described herein include, as appropriate: energy sources (e.g., geothermal aquifer, reservoir, oil/gas field); energy production facilities (e.g., well, platform, dam, pump); fuel delivery systems (e.g., offshore or overland pipeline, barge, tanker, highway transport, belt conveyor, aerial tramway); power generating systems (e.g., turbine, generator); transmission systems (e.g., right-of-way, switchyard, substation); pollution control systems (e.g., drilling muds and cuttings, stack gas emission control, non-point source emission control, cooling water and wastewater treatment and discharge, ash disposal); supplies (e.g., location of stocks of parts and chemicals, transport routes); staffing (e.g., numbers of workers, skill requirements); services (e.g., fire protection, security, transportation); planning for emergencies, and community involvement (e.g., worker housing during construction).

3. <u>Objectives</u>. This section will summarize the general scope of the environmental assessment and discuss its timing in relation to other aspects of project preparation, design, and execution. This section will identify constraints, if any, regarding the adequacy of existing environmental assessment baseline data and needs to phase additional data collection (e.g., over several seasons) and assessment efforts so as not to hinder the rest of the project development schedule.

4. <u>Environmental Assessment Requirements</u>. This section will identify regulations and guidelines which will govern the conduct of the assessment or specify the content of its report. They may include any or all of the following:

- World Bank Operational Directive 4.00, Annex A: "Environmental Assessment," and other pertinent ODs, OMSs, OPNs, and guidelines;
- national laws and/or regulations on environmental reviews and impact assessments;
- regional, provincial or communal environmental assessment regulations; and
- environmental assessment regulations of any other financing organizations involved in the project.

This section will identify design or operating standards which project components must address to be environmentally acceptable. This will include, for example, effluent discharge limitations, air emission standards, receiving water quality standards, and occupational health and safety requirements.

5. <u>Study Area</u>. This section will specify the boundaries of the study area for the assessment. Where appropriate, specify the right-of-way (ROW) width and alignment for transmission lines or pipelines. Similarly, specify locations for transmission substations, oil/gas compressor or pump stations. For projects which develop energy sources, specify the entire area involved (e.g., catchment and floodplain for hydroelectric reservoirs and the production and reserve zones for oil/gas fields).

If there are adjacent or remote areas which should be considered with respect to impacts of particular aspects of the project, identify them. For example, when an energy project includes only a thermoelectric power plant and does not include the oil/gas development component, transportation corridors, terminals, and remote processing locations for fuel delivery should be identified.

6. <u>Scope of Work</u>. In some cases, the tasks to be carried out by a consultant will be known with sufficient certainty to be specified completely in the terms of reference. In other cases, information deficiencies need to be alleviated or specialized field studies or modelling activities performed to assess impacts, and the consultant will be asked to define particular tasks in more detail for contracting agency review and approval.

7. <u>Task 1. Describe the proposed project</u>. Provide information on the following: location of all project-related development sites and ROW's; general layout of facilities at project-related development sites; flow diagrams of facilities/operations design basis, size, capacity, flow-through of unit operations; pre-construction activities; construction activities, schedule, staffing and support, facilities and services; operation and maintenance activities, staffing and support, facilities and services; required off-site investments; life expectancy for major components.

Provide maps at appropriate scales to illustrate the general setting of project-related development sites and ROW's, as well as surrounding areas likely to be environmentally affected. These maps shall include topographic contours, as available, as well as locations of major surface waters, roads, railways, town centers, parks and preserves, and political boundaries. Also provide, as available, maps to illustrate existing land use, including industrial, residential, commercial and institutional development, agriculture, etc.

8. <u>Task 2. Description of the Environment</u>. Assemble, evaluate and present baseline data on the environmental characteristics of the study area. Include information on any changes anticipated before the project commences.

   Physical environment: geology (e.g., stratigraphy and structure of well fields, seismic history of storage tank areas); topography (e.g., drainage patterns around construction areas, view-sheds around facilities); soils (e.g., agricultural value, potential use for lining, or soil cover in residue disposal); benthic sediment (e.g., level of contamination in offshore areas where pipelines are laid); climate and meteorology (e.g., prevailing wind patterns around stacks, precipitation patterns at residue disposal sites); ambient air quality (note input from other major pollutant generators in the area, if any); surface water hydrology (e.g., downstream water resources from reservoirs); coastal and oceanic parameters (e.g., currents in platform and docking areas); receiving water quality (note input from major pollutant generators in the area, if any); significant pollutant sources in the area and prospects for their mitigation.

   Biological environment: flora (e.g., types and diversity); fauna (e.g., resident and migratory); rare or endangered species within or in areas adjacent to project-related development sites or ROW's; sensitive habitats, including wetlands, parks or preserves, significant wildlands within or in areas downstream/downgradient of project-related development areas or ROW's (including benthic habitat in areas of offshore pipelines); species of commercial importance in areas affected by the project, including coastal areas at docking facilities.

   Socio-cultural environment (include both present and projected where appropriate): population (i.e., full time and seasonal); land use (i.e., year-round and seasonal); planned development activities; community structure; employment and labor market; distribution of income, goods and services; recreation; public health; education; cultural properties (e.g., archaeological and historically significant sites); indigenous peoples and traditional tribal lands; customs, aspirations and attitudes.

9. <u>Task 3. Legislative and Regulatory Considerations</u>. Describe the pertinent regulations and standards governing environmental quality, health and safety, protection of sensitive areas, protection of endangered species, siting, land use control, etc., at international, national, regional and local levels. (The TOR should specify those that are known and require the consultant to investigate for others.)

10. <u>Task 4. Determination of the Potential Impacts of the Proposed Project</u>. Identify all significant changes which the project would incur. These would include, but not be limited to, changes in the following: employment opportunities, wastewater effluents, thermal effluents, air emissions, land use, infrastructure, exposure to disease, noise, traffic, socio-cultural behavior. Assess the impacts from changes brought about by the project on baseline environmental conditions as described above under Task 2.

   In this analysis, distinguish between significant positive and negative impacts, direct and indirect impacts, and immediate and long-term impacts. Include indirect impacts from the increased power supply (e.g., industrial expansion and increased urbanization). Identify impacts which may occur due to accidental events (e.g., potential rupture of oil pipelines, leakage from a gas pipeline, blow-out of an oil well, tanker collision). Identify impacts which are unavoidable or irreversible.

Wherever possible, describe impacts quantitatively, in terms of environmental costs and benefits. Assign economic values when feasible.

Impact analysis for energy projects should be divided between construction impacts and operation impacts. For example, for pipelines there are construction impacts of land clearing (e.g., loss of vegetative habitat for wildlife) and operation impacts of pipeline maintenance (e.g., use of herbicides). For thermoelectric power plants, there are construction impacts of housing construction workers (e.g., market demand changes for local services) and operation impacts of power plant operation (e.g., stack gas emissions and effluent discharges).

Characterize the extent and quality of available data, explaining significant information deficiencies and any uncertainties associated with predictions of impact. If possible, give the TOR for studies to obtain the missing information. For information which could not be obtained until after project execution, provide TOR for studies to monitor operations over a given time period and to modify designs and/or operational parameters based upon updated impact analysis.

11. Task 5. Analysis of Alternatives to the Proposed Project. The environmental assessment should include an analysis of reasonable alternatives to meet the ultimate project objective. This analysis may suggest designs that are more sound from an environmental, sociocultural or economic point of view than the originally proposed project. Include the "no action" alternative -- not constructing the project -- in order to demonstrate environmental conditions without it. Alternatives should include the following: the "no action" alternative (as discussed above); alternative means of meeting the energy requirements; the alternative of upgrading existing facilities; alternative routes and sites; alternative design; and alternative methods of construction, including costs and reliability.

Describe how the alternatives compare in terms of potential environmental impacts; capital and operating costs; suitability under local conditions (e.g., skill requirements, political acceptability, public cooperation, availability of parts, level of technology); and institutional, training, and monitoring requirements. When describing the impacts of alternatives, indicate which impacts would be irreversible or unavoidable and which could be mitigated.

To the extent possible, quantify the costs and benefits of each alternative, incorporating the estimated costs of any associated mitigating measures. Describe the reasons for selecting the proposed project over the other alternatives.

12. Task 6. Development of Management Plan to Mitigate Negative Impacts. For the proposed project, recommend feasible and cost-effective measures to prevent or reduce significant negative impacts to acceptable levels. Include measures to address emergency response requirements for accidental events.

Estimate the impacts and costs of those measures, and of the institutional and training requirements to implement them. Consider compensation to affected parties for impacts which cannot be mitigated. Prepare a management plan including proposed work programs, budget estimates, schedules, staffing and training requirements, and other necessary support services to implement the mitigating measures.

13. Task 7. Identification of Institutional Needs to Implement Environmental Assessment Recommendations. Review the authority and capability of institutions at local, provincial/regional, and national levels and recommend steps to strengthen or expand them so that the management and monitoring plans in the environmental assessment are likely to be implemented. The recommendations may extend to new laws and regulations, new agencies or agency functions, intersectoral arrangements, management procedures and training, staffing, operation and maintenance training, budgeting, and financial support.

14. Task 8. Development of a Monitoring Plan. Prepare a detailed plan to monitor the implementation of mitigating measures and the impacts of the project during construction and operation. Include in the plan an estimate of capital and operating costs and a description of other inputs (such as training and institutional strengthening) needed to conduct it.

15. Task 9. Assist in Inter-Agency Coordination and Public/NGO Participation. Assist in coordinating the environmental assessment with other government agencies, in obtaining the views of local NGO's and affected groups, and in keeping records of meetings and other activities, communications, and comments and their disposition. (The TOR should specify the types of activities; for example, interagency scoping session, environmental briefings for project staff and interagency committees, support to environmental advisory panels, public forum.)

16. Report. Provide an environmental assessment report which is concise and limited to significant environmental issues. The main text should focus on findings, conclusions and recommended actions, supported by summaries of the data collected and citations for any references used in interpreting those data. Detailed or uninterpreted data are not appropriate in the main text and should be presented in appendices or a separate volume. Unpublished documents used in the assessment may not be readily available and should also be assembled in an appendix. Organize the environmental assessment report according to the outline below. (This is the format suggested in OD 4.00, Annex A-1; the TOR may specify a different one to satisfy national agency requirements as long as the topics required in the Bank's directive are covered):

    - Executive Summary
    - Policy, Legal and Administrative Framework
    - Description of the Proposed Project
    - Description of the Environment
    - Significant Environmental Impacts
    - Analysis of Alternatives
    - Mitigation Management Plan
    - Environmental Management and Training
    - Monitoring Plan
    - Inter-Agency and Public/NGO Involvement
    - List of References
    - Appendices:
        - List of Environmental Assessment Preparers
        - Records of Inter-Agency and Public/NGO Communications
        - Data and Unpublished Reference Documents

17. <u>Consulting Team</u>. The environmental assessment requires interdisciplinary analysis. The general skills required of an environmental assessment team are: environmental management planning, socio-economics, ecology, hydrology/hydrogeology, air quality analysis, water quality analysis. For an energy project, the project team will be specified to include specialists appropriate to the type of components in the energy project (e.g., for offshore pipelines, an oceanographer and marine biologist; for transmission lines, a terrestrial biologist and cultural resources specialist; for thermoelectric power plants, an air quality modeler and aquatic biologist; for hydroelectric projects, a hydrologist and aquatic biologist). When possible, the TOR should provide an estimate of staff weeks/months required.

18. <u>Schedule</u>. This section will specify dates for progress reviews, interim and final reports, and other significant events.

19. <u>Other Information</u>. Include here lists of data sources, project background reports and studies, relevant publications, and other items to which the consultant's attention should be directed.

ANNEX 10-2

Sample Terms of Reference (TOR)
An Environmental Assessment of Industrial Facilities

Note: Paragraph numbers correspond to those in the
Sample Terms of Reference (TOR) Outline in
Annex 1-3; additional paragraphs are not numbered

1. <u>Introduction</u>. This section will state the purpose of the terms of reference, identify the industrial development project to be assessed, and explain the executing arrangements for the environmental assessment. Industrial development projects include, but are not limited to: industrial production facilities (e.g., chemical, petrochemical, pulp and paper, iron and steel, nonferrous metals, petroleum refining, cement, fertilizer, and food processing plants); raw materials sources (e.g., mines and wells, and related handling, processing and storage facilities); raw materials and product transportation facilities (e.g., marine terminals, deepwater ports, pipelines, roads, rail); and industrial pollution control facilities (e.g., waste minimization systems, hazard reduction and emergency response systems, air emission control, wastewater treatment, residual disposal).

2. <u>Background Information</u>. This section will provide pertinent background for potential parties who may conduct the environmental assessment, whether they are consultants or government agencies. The section will include a brief description of the major components of the proposed project, a statement of the need for it and the objectives it is intended to meet, the implementing agency, a brief history of the project (including alternatives considered), its current status and timetable, and the identities of any associated projects. If there are other projects in progress or planned within the region which may compete for the same resources, identify them within this section.

   Major components of an industrial project to be described herein include, as appropriate: local and foreign raw material sources (e.g., hard rock mines, oil/gas wells, chemical plants, slaughterhouses, produce farms); processing operations (e.g., process flow sequence, continuous or batch, size, production); expected markets for products (e.g., local and foreign markets); transport systems (e.g., roads, pipelines, rail, barge); pollution control systems (e.g., source reduction and recycling to minimize wastes, stack gas emission control, non-point source emission control, wastewater treatment and discharge, solid waste disposal, spill prevention); supplies (e.g., location of stocks of parts and chemicals, transport routes); staffing (e.g., numbers of workers, skill requirements); services (e.g., fire protection, security, transportation, medical); and community involvement (e.g., worker housing during construction).

3. <u>Objectives</u>. This section will summarize the general scope of the environmental assessment and discuss its timing in relation to other aspects of project preparation, design, and execution. This section will identify constraints, if any, regarding the adequacy of existing environmental assessment baseline data and needs to phase additional data collection (e.g., over several seasons) and assessment efforts so as not to hinder the rest of the project development schedule.

4. Environmental Assessment Requirements. This section will identify regulations and guidelines which will govern the conduct of the assessment or specify the content of its report. They may include any or all of the following:

- World Bank Operational Directive 4.00, Annex A: "Environmental Assessment," and other pertinent ODs, OMSs, OPNs, and Guidelines;
- national laws and/or regulations on environmental reviews and impact assessments;
- regional, provincial or communal environmental assessment regulations; and
- environmental assessment regulations of any other financing organizations involved in the project.

This section will identify design or operating standards which project components must address to be environmentally acceptable. This will include, for example, effluent discharge limitations, air emission standards, receiving water quality standards, and occupational health and safety requirements.

5. Study Area. This section will specify the boundaries of the study area for the assessment. Where appropriate, specify the right-of-way (ROW) width and alignment for pipelines and transportation corridors for raw material and product shipments. For projects including mines and oil/gas wells, include the boundaries of the related ore bodies and well fields, respectively.

If there are adjacent or remote areas which should be considered with respect to impacts of particular aspects of the project, identify them. For example, where intermediate supplies for a processing operation will be generated at remote facilities, identify the remote facilities (e.g., identify the sources of intermediate chemical supplies to be used at a pharaceutical plant), because an added demand for supplies from this remote facility may cause an environmental impact to the remote area.

6. Scope of Work. In some cases, the tasks to be carried out by a consultant will be known with sufficient certainty to be specified completely in the terms of reference. In other cases, information deficiencies need to be identified and alleviated or specialized field studies or modelling activities performed to assess impacts, and the consultant will be asked to define particular tasks in more detail for contracting agency review and approval.

7. Task 1. Describe the proposed project. Provide information on the following: location of all project-related development sites and ROW's; general layout of facilities at project-related development sites; flow diagrams of facilities/operations; design basis, size, capacity, flow-through of unit operations; pre-construction activities; construction activities, schedule, staffing and support, facilities and services; operation and maintenance activities, staffing and support, facilities and services; reclamation activities, such as in mining projects; required off-site investments; life expectancy for major components.

Provide maps at appropriate scales to illustrate the general setting of project-related development sites and ROW's, as well as surrounding areas likely to be environmentally affected. These maps shall include topographic contours, as available, as well as locations of major surface waters, roads, railways, town centers, parks and preserves, and political boundaries. Also provide, as available, maps to illustrate existing land uses.

8. <u>Task 2. Description of the Environment</u>. Assemble, evaluate and present baseline data on the environmental characteristics of the study area. Include information on any changes anticipated before the project commences.

Physical environment: geology (e.g., stratigraphy and structure of well fields, seismic history of storage tank areas, integrity of geological layers protecting potable groundwater supplies); topography (e.g., drainage patterns around construction areas, view-sheds around facilities); soils (e.g., agricultural value, potential use for lining or soil cover in residue disposal); climate and meteorology (e.g., prevailing wind patterns around stacks, precipitation patterns at residue disposal sites); ambient air quality (e.g., ability to assimilate emissions and maintain air quality standards); (note input from other major pollutant generators in the area, if any); surface water hydrology (e.g., downstream water resources from reservoirs, soil erosion and sedimentation potential, flood hazard potential); water resources (e.g., adequacy of water supplies); coastal and oceanic parameters (e.g., currents in docking areas, dispersion potential at effluent discharge locations); receiving water quality (e.g., ability to assimilate effluent discharges and maintain water quality standards for desired uses); (note input from major pollutant generators in the area, if any); significant pollutant sources in the area and prospect for their mitigation.

Biological environment: flora and fauna; rare or endangered species within or in areas adjacent to project-related development sites or ROW's; sensitive habitats, including wetlands, parks or preserves, significant wildlands within or in areas downstream/downgradient of project-related development areas or ROW's; species of commercial importance in areas affected by the project, including coastal areas at docking facilities.

Socio-cultural environment (include both present and projected where appropriate): population (i.e., full time and seasonal); land use (i.e., year-round and seasonal); planned development activities; community structure; employment and labor market; distribution of income, goods and services; recreation; public health; education; cultural properties (e.g., archaeological and historically significant sites); indigenous peoples and traditional tribal lands; customs, aspirations and attitudes.

9. <u>Task 3. Legislative and Regulatory Considerations</u>. Describe the pertinent regulations and standards governing environmental quality, health and safety, protection of sensitive areas, protection of endangered species, siting, land use control, etc., at international, national, regional and local levels. (The TOR should specify those that are known and require the consultant to investigate for others.)

10. <u>Task 4. Determination of the Potential Impacts of the Proposed Project</u>. Identify all significant changes which the project would incur. These would include, but not be limited to, changes in the

following: employment opportunities, wastewater effluents, air emissions, solid wastes, land use, infrastructure, exposure to disease, risk of industrial hazard, noise, traffic, sociocultural behavior. Assess the impacts from changes brought about by the project on baseline environmental conditions as described above under Task 2.

In this analysis, distinguish between significant positive and negative impacts, direct and indirect impacts, and immediate and long-term impacts. Identify impacts which are unavoidable or irreversible. Wherever possible, describe impacts quantitatively, in terms of environmental costs and benefits. Assign economic values when feasible.

Impact analysis for industrial projects should be divided between construction impacts and operation impacts. For example, for pipelines, there are potential construction impacts of land clearing (e.g., loss of vegetative habitat for wildlife) and operation impacts of pipeline maintenance (e.g., use of herbicides). For mines, there are potential construction impacts of land clearing (e.g., loss of lands for other uses, such as agriculture), operation impacts of materials handling (e.g., dusts from mining and milling, tailings disposal), and reclamation (e.g., return of the land to a natural state). For industrial manufacturing plants, there are potential construction impacts of housing construction workers (e.g., market demand changes for local services) and operation impacts from process operations (e.g., stack emissions, effluent discharges, noises, industrial hazards).

Assess the risk of occurence of potential industrial hazards (e.g., accidental spills, fires, explosions, impoundment structural failure, gaseous releases). Consider the ability of the community to provide emergency response services for potential industrial hazards. Consider the ability of the community to provide medical services to respond to emergencies. Based on the above, assess the potential impacts.

Characterize the extent and quality of available data, explaining significant information deficiencies and any uncertainties associated with predictions of impact. If possible, give the TOR for studies to obtain the missing information. For information which could not be obtained until after project execution commences, provide TOR for studies to monitor operations over a given time period and to modify designs and/or operational parameters based upon updated impact analysis.

11. **Task 5. Analysis of Alternatives to the Proposed Project**. The environmental assessment should include an analysis of reasonable alternatives to meet the ultimate project objective. The analysis may lead to designs that are more sound from an environmental, sociocultural or economic point of view than the original project proposal. The concept of alternatives extends to siting, design, fuels, raw materials and technology selection, construction techniques and phasing, and operating and maintenance procedures. Include the "no action" alternative -- not constructing the project -- in order to demonstrate environmental conditions without it. Alternatives should include the following: the "no action" alternative (as discussed above); alternative means of meeting industrial product requirements; the alternative of upgrading existing facilities; alternative routes and sites; alternative design; and alternative methods of construction, including costs and reliability.

Describe how the alternatives compare in terms of potential environmental impacts; capital and operating costs; suitability under local conditions (e.g., skill requirements, political acceptability, public cooperation, availability of parts, level of technology); and institutional, training, and monitoring requirements. When describing the impacts of alternatives, indicate which impacts would be irreversible or unavoidable and which could be mitigated.

To the extent possible, quantify the costs and benefits of each alternative, incorporating the estimated costs of any associated mitigating measures. Describe the reasons for selecting the proposed project over the other alternatives.

12. Task 6. Development of Management Plan to Mitigate Negative Impacts. For the proposed project, recommend feasible and cost-effective measures to prevent or reduce significant negative impacts to acceptable levels. Include measures for emergency response to accidental events (e.g., ruptures, leaks, tanker truck or ship accidents, fires, explosions), as appropriate. Estimate the impacts and costs of those measures, and of the institutional and training requirements to implement them. Consider compensation to affected parties for impacts which cannot be mitigated. Prepare a management plan including proposed work programs, budget estimates, schedules, staffing and training requirements, and other necessary support services to implement the mitigating measures.

13. Task 7. Identification of Institutional Needs to Implement Environmental Assessment Recommendations. Review the authority and capability of institutions at local, provincial/regional, and national levels and recommend steps to strengthen or expand them so that the management and monitoring plans in the environmental assessment are likely to be effectively implemented. The recommendations may extend to new laws and regulations, new agencies or agency functions, intersectoral arrangements, management procedures and training, staffing, operation and maintenance training, budgeting, and financial support.

14. Task 8. Development of a Monitoring Plan. Prepare a detailed plan to monitor the implementation of mitigating measures and the impacts of the project during construction and operation. Include in the plan an estimate of capital and operating costs and a description of other inputs (such as training and institutional strengthening) needed to conduct it.

15. Task 9. Assist in Inter-Agency Coordination and Public/NGO Participation. Assist in coordinating the environmental assessment with other government agencies, in obtaining the views of local NGO's and affected groups, and in keeping records of meetings and other activities, communications, and comments and their disposition. (The TOR should specify the types of activities; e.g., interagency scoping session, environmental briefings for project staff and interagency committees, support to environmental advisory panels, public forum.)

16. Report. Provide an environmental assessment report which is concise and limited to significant environmental issues. The main text should focus on findings, conclusions and recommended actions, supported by summaries of the data collected and citations for any references used in interpreting those data. Detailed or uninterpreted data are not appropriate in the main text and should

be presented in appendices or a separate volume. Unpublished documents used in the assessment may not be readily available and should also be assembled in an appendix. Organize the environmental assessment report according to the outline below. (This is the format suggested in OD 4.00, Annex A-1; the TOR may specify a different one to satisfy national agency requirements as long as the topics required in the Bank's directive are covered):

- Executive Summary
- Policy, Legal and Administrative Framework
- Description of the Proposed Project
- Description of the Environment
- Significant Environmental Impacts
- Analysis of Alternatives
- Mitigation Management Plan
- Environmental Management and Training
- Monitoring Plan
- Inter-Agency and Public/NGO Involvement
- List of References
- Appendices:
  - List of Environmental Assessment Preparers
  - Records of Inter-Agency and Public/NGO Communications
  - Data and Unpublished Reference Documents

17. <u>Consulting Team</u>. The environmental assessment requires interdisciplinary analysis. The general skills required of an environmental assessment team are: environmental management planning, ecology, socio-economics, ecology, hydrology/hydrogeology, air quality analysis, water quality analysis, transportation planning. For an industrial project, the project team will be specified to include specialists appropriate to the type of components in the industrial project as needed (e.g., for deepwater ports, an oceanographer and marine biologist; for pipelines, a terrestrial biologist and cultural resources specialist; for industrial manufacturing plants, an industrial process engineer, and air quality specialist; for industrial wastewater treatment, a civil/sanitary engineer and aquatic biologist). When possible, the TOR should provide an estimate of staff weeks/months required.

18. <u>Schedule</u>. This section will specify dates for progress reviews, interim and final reports, and other significant events.

19. <u>Other Information</u>. Include here lists of data sources, project background reports and studies, relevant publications, and other items to which the consultant's attention should be directed.

ABBREVIATIONS/ACRONYMS

ACC/SCN	(U.N.) Administrative Committee on Co-ordination/Sub Committee on Nutrition
ADB	African Development Bank (see AfDB) and Asian Development Bank (see AsDB)
AfDB	African Development Bank (see ADB)
AGR	Sociological Advisor
AICE	American Institute of Chemical Engineers
AID	(United States) Agency for International Development (see USAID)
API	American Petroleum Institute
AsDB	Asian Development Bank (see ADB)
ASEAN	Association of Southeast Asian Nations
ATL	Action Threshold Level
BOD	Biochemical Oxygen Demand
BOD_5	Biochemical Oxygen Demand Over Five Days
BOOT	Build-Own-Operate-Transfer
BOT	Build-Operate-Transfer
BTO	Back-To-Office Report
C	Carbon
°C	(Centigrade) Celsius
CBA	Cost-Benefit Analysis
CD	Country Department
CECC	Credit Extension Coordinating Committee
CFC	Chlorofluorocarbons
CFR	Code of Federal Regulations
CGIAR	Consultative Group on International Agricultural Research
CH_4	Methane
CIDA	Canadian International Development Agency
CITES	(UN) Convention on International Trade in Wild Flora and Fauna
CO	Carbon monoxide
CO_2	Carbon Dioxide
COD	Chemical Oxygen Demand
	Country Operations Division
CMEA	Council for Mutual Economic Assistance
CSI	Construction Specifications Institute
CSP	Country Strategy Paper
DANIDA	Danish International Development Agency
DAP	diammonium phosphate
dB	decibel
DCCI	Development Commission for Cement Industry (India)
DFI	Development Finance Institution
DMG	Drylands Management Guidelines
DNA	Deoxyribonucleic Acid
EA	Environmental Assessment
EA OD	Environmental Assessment Operational Directive

EAPs	Environmental Action Plans
EAR	Environmental Assessment Report
EBRD	European Bank for Reconstruction and Development
EC	European Communities
ECU	European currency unit
EDF	European Development Fund
EDI	Economic Development Institute of the World Bank
EDP	Environmentally-adjusted Net Domestic Product
EEC	European Economic Community
EIB	European Investment Bank
EIP	Environmental Issues Paper
EMF	Electro Magnetic Field
ENV	Environment Department
EPA	(United States) Environmental Protection Agency (see USEPA)
EPD	Environmental Protection Department
EPS	Executive Project Summary
ER	Environmental Review
ERL	Emergency Recovery Loan
	Emergency Reconstruction Loan
ERR	Economic Rate of Return
ESCAP	Economic and Social Commission for Asia and the Pacific
ESMAP	Energy Sector Management Assistance Program
EXTIE	International Economic Relations Division of the External Affairs Department (World Bank)
°F	Fahrenheit
FAO	Food and Agricultural Organization of the United Nations
FEPA	(Nigerian) Federal Environmental Protection Agency
FEPS	Final Executive Project Summary
FI	Financial Intermediary
FIL	Financial Intermediary Loan
FINNIDA	Finnish Department of International Development Cooperation
FMWH	Federal Ministry of Works and Housing
FY	Fiscal Year
GATT	(UN) General Agreement on Tariffs and Trade
GDP	Gross Domestic Product
GEF	Global Environment Facility
GLC	ground level concentrations
GNP	Gross National Product
GOI	Government of India
GOR	Government of Rwanda
GTZ	German Agency for Technical Cooperation
ha	hectare
HABITAT	United Nations Centre for Human Settlements
HHS	(United States) Department of Health and Human Services

H_2S	Hydrogen Sulfide
IADB	Inter-American Development Bank
IBRD	International Bank for Reconstruction and Development
ICICI	Industrial Credit and Investment Corporation of India Limited
ICCROM	International Centre for the Study of the Preservation and the Restoration of Cultural Property
ICOMOS	International Committee of Monuments and Sites
IDA	International Development Association
IDB	Inter-American Development Bank
IDBI	Industrial Development Bank of India
IEC	information, education, and communication
IEPS	Initial Executive Project Summary
IFAD	(UN) International Fund for Agricultural Development
IFC	International Finance Corporation of the World Bank
IIED	International Institute for Economic Development
ILO	(UN) International Labour Organization
IMF	International Monetary Fund
IMO	International Maritime Organization
IPCC	Intergovernmental Panel on Climate Change
IPM	Integrated Pest Management
IRR	Internal Rate of Return
ITTO	International Tropical Timber Organization
IUCN	International Union for Conservation of Nature and Natural Resources
km	kilometer
km^2	square kilometer
KV	kilovolts
kW	kilowatts
m	meter
MAP	monoammonium phosphate
mm	millimeter
MIGA	Multilateral Investment Guarantee Agency of the World Bank
Miniplan	Ministry of Plan
MOE	Ministry of Environmental Protection
MOP	Memorandum of the President
MOS	Monthly Operational Summary
MW	Mega Watts
NEAP	National Environmental Action Plan
NEC	National Environment Commission
NGO	Nongovernmental Organization
NIOSH	National Institute for Occupational Safety and Health
NORAD	Norwegian Agency for International Development
NO_x	Oxides of Nitrogen
NPV	Net Present Value
N_2O	Nitrous Oxide

O, O_2, O_3	Oxygen
OAU	Organization for African Unity
OD	Operational Directive
ODA	Overseas Development Administration (United Kingdom)
OECD	Organization for Economic Cooperation and Development
OED	Operations Evaluation Department
O/G	oil and grease
OIEC	Organization for International Cooperation
OMS	Operational Manual Statement
OPN	Operations Policy Note
OPNSV	Senior Vice President, Operations
PB	Project Brief
PCB	polychlorinated biphenyl
PCR	Project Completion Report
PEPA	Pakistan Environmental Protection Agency
pH	measurement of acidity and alkalinity (on log. scale 0-14, 7 = neutral, ≤ 7 = increasing acidity, ≥ 7 = increasing alkalinity)
PHN	Population, Health and Nutrition Department (World Bank)
PIDs	Provincial Irrigation Departments
PI/ER	Public Investment/Expenditure Review
PIP	Public Investment Program
PIR	Project Implementation Review
POPTR	Personnel Operations Department, Training Division
ppb	parts per billion
PPF	Project Preparation Facility
PPR	Project Performance Report
PR	President's Report
PRC	People' Republic of China
PRE	Policy, Research, and External Affairs
R&D	Research and Development
RED	Regional Environment Division
ROW	Right-of-Way
RVP	Regional Vice President
SAL	Structural Adjustment Lending
	Structural Adjustment Loan
SAR	Staff Appraisal Report
SCBA	Social Cost Benefit Analysis
SDC	Swiss Development Corporation
SECAL	Sector Adjustment Loan
SIDA	Swedish International Development Authority
SNA	System of National Accounts
SOD	Sector Operations Division
SO_2	Sulfur Dioxide
SO_x	Oxides of Sulfur

SPPF	Special Project Preparation Facility
SPRIE	International Economic Relations Division of the Strategic Planning Department
STEL	Short-Term Exposure Limit
TAL	Technical Assistance Loan
TDS	Total Dissolved Solids
TFAP	Tropical Forest Action Plan
TLV	Threshold Limit Values
TM	Task Manager
TMA	time weighted average
TOC	Total Organic Carbon
TOR	Terms of Reference
TSD	Toxic Storage Disposal
TSP	Total Suspended Particulates
TSS	Total Suspended Solids
UNDP	United Nations Development Programme
UNDRO	United Nations Disaster Relief Organization
UNEP	United Nations Environment Programme
UNESCO	United Nations Educational, Scientific and Cultural Organization
UNICEF	United Nations Children's Fund
UNIDO	United Nations Industrial Development Organization
UNSO	United Nations Statistical Office
USACE	United States Army Corps of Engineers
USAID	United States Agency for International Development (see AID)
USEPA	United States Environmental Protection Agency (see EPA)
WAPDA	Water and Power Development Authority
WHO	World Health Organization
WMA	Wildlands Management Area
WUAs	Water User Associations

ENVIRONMENTAL ASSESSMENT

A Guide To Further Reading

Abel, N., and M. Stocking. 1981. "The Environmental Assessment Experience of Underdeveloped Countries." In Project Appraisal and Policy Review, edited by T. O'Riordan and W. R. D. Sewell. Chichester, United Kingdom: John Wiley and Sons.

___. 1990. Environmental Risk Assessment: Dealing with Uncertainty in Environmental Impact Assessment. Environment Paper 7. Manila, Philippines: Office of the Environment.

Ahmad, Y.J., and G. K. Sammy. 1985. Guide to Environmental Impact Assessment in Developing Countries. London: Hodder and Stoughton (for the United Nations Environment Programme).

___. 1987. Orientaciones para la Evaluacion del Impacto Ambiental en los Paises en Desarrollo. Nairobi, Kenya: PNUMA (also in French).

American Arbitration Association. 1980. Improving EIS Scoping. Washington D.C.

American Society of Civil Engineers. 1989. Guidelines for Planning and Designing Hydroelectric Developments. Volume I: Planning, Design of Dams and Related Topics, and Environmental. New York: ASCE.

___. 1989. Guidelines for Planning and Designing Hydroelectric Developments. Volume II: Waterways. New York: ASCE.

Anderson, A., ed. 1990. Alternatives to Deforestation. New York: Columbia University Press.

Andrews, R. N. L., and others. 1977. Substantive Guidance for Environmental Impact Assessment: An Exploratory Study. Indianapolis, Indiana: Butler University, Holcomb Research Institute and the Institute of Ecology.

Anon, 1988. "The International Development of Environmental Impact Assessment." The Environmentalist 8(2):143.

___. 1988. "Transportation Elements of Environmental Impact Assessments and Reports." Institute of Transportation Engineering Journal 58(6):69-76.

Asian Development Bank. 1986. Environmental Guidelines for Selected Infrastructure Projects. 1 Volume. Manila, Philippines: Infrastructure Department, Environment Unit.

___. 1988e. Guidelines for Integrated Regional Economic-Environmental Development Planning: A Review of Regional Environmental Development Planning Studies in Asia. Paper No. 3, Volume I: Guidelines. Manila, Philippines.

____. 1988f. Guidelines for Integrated Regional-Environmental Development Planning: A Review of Regional Environmental Development Planning Studies in Asia. Paper No. 3, Volume II: Case Studies. Manila, Philippines.

____. 1987. Handbook on the Use of Pesticides in the Asia-Pacific Region. Manila, Phillipines.

Association of Southeast Asian Nations. 1985. Report of Workshop on the Evaluation of Environmental Impact Assessment Applications in ASEAN countries. Bandung, Indonesia.

Ayanda, J. O. 1988. "Incorporating Environmental Impact Assessment in the Nigerian Planning Process: Need and Procedure." Third World Planning Review 10:51-64 (U.K.).

Barbier, E. B. 1990. "Alternative Approaches to Economic-Environmental Interactions." Ecology Economics 2:7-26.

____. 1989. Economics, Natural Resource Scarcity and Development: Conventional and Alternative Views. London: Earthscan Publications Ltd.

____. 1988. "Economic Valuation of Environmental Impacts." Project Appraisal 3:143-150.

____. 1991. "Environmental Sustainability and Cost-Benefit Analysis." Environment and Planning 22:1259-1266.

Barbier, E. B., A. Markandya, and D. W. Pearce. 1990. "Sustainable Agricultural Development and Project Appraisal." European Review of Agrarian Economics 17(2): 181-196.

Barrett, B. P. D., and R. Therivel. 1990. Environmental Policy and Impact Assessment in Japan. London, United Kingdom: Routledge (in press).

Bartlett, R. V., ed. 1989. Policy Through Impact Assessment: Institutionalized Analysis as a Policy Strategy. New York: Greenwood.

Bauchum, R. G. 1985. Needs Assessment Methodologies in the Development of Impact Statements. Monticello, Illinois: Vance Bibliographies.

Becker, D.S., and J. W. Armstrong. 1988. "Development of Regionally Standardized Protocols for Marine Environmental Studies." Marine Pollution Bulletin 19(7):310-313.

Becker, H. A., and A. L. Porter, eds. 1986. Methods and Experiences in Impact Assessment. Atlanta, Georgia: International Association for Impact Assessment.

Bellia, V., and E. D. Bidone. 1990. Rodovias, Recursos Naturais e Meio Ambiente. Rio de Janerio: Departamento Nacional de Estradas e Rodagem.

Bendix, S., and H. R. Graham, eds. 1986. Environmental Assessment: Approaching Maturity. Ann Arbor, Michigan: Ann Arbor Science.

Bisset, R. 1980. "Methods for Environmental Impact Analysis: Recent Trends and Future Prospects." Journal of Environmental Management 11:27-43.

___. 1984. "Post Development Audits to Investigate the Accuracy of Environmental Impact Predictions." Zeitschrift fur Umweltpolitik 7:463-484.

Biswas, A. K., and Qu Geping [Chu, Ko-Ping], eds. 1987. Environmental Impact Assessment for Developing Countries. Oxford, United Kingdom: Tycooly International (for the United Nations University).

Bochniarz, Z., and A. Kassenberg. 1985. Environmental Protection by Integrated Planning. Warsaw, Poland: Economic and Social Problems of Environmental Planning and Processing.

Bojo, J., K-G Maler, and L. Unemo. 1988. Economic Analysis of Environmental Consequences of Development Projects. Stockholm, Sweden: The Economic Research Institute, Stockholm School of Economics.

Bowden, M-A, and F. Curtis. 1988. "Federal EIA in Canada: EARP as an Evolving Process." Environmental Impact Assessment Review 8(1): 97-106.

Bowonder, B., and S. S. Arvind. 1989. "Environmental Regulations and Litigation in India." Project Appraisal 4:182-196.

Bregha, F. and others. 1990. The Integration of Environmental Considerations into Government Policy. Hull, Quebec: The Rawson Academy of Aquatic Science.

Burkhardt, D. F., and W. H. Ittelson, eds. 1978. Environmental Assessment of Socioeconomic Systems. New York: Plenum.

Cable, T. T., V. Brack, and V. R. Holmes. 1989. "Simplified Method for Wetland Habitat Assessment." Environmental Management 13:207-13.

Cairns, J., and T. V. Crawford, eds. 1991. Integrated Environmental Management. Chelsea, Michigan: Lewis Publishers.

Campbell, M. J. 1990. New Technology and Rural Development: The Social Impact. London, United Kingdom: Routledge.

Canter, L. W. 1977. Environmental Impact Assessment. New York: McGraw-Hill.

___. 1985. Environmental Impact of Water Resources Projects. Chelsea, Michigan: Lewis Publishers.

___. 1982. <u>The Status of Environmental Impact Assessment in Developing Countries</u>. Norman: University of Oklahoma, Environment and Ground Water Institute.

Canter, L. W., and L. G. Hill. 1981. <u>Handbook of Variables for Environmental Impact Assessment</u>. Ann Arbor, Michigan: Ann Arbor Science.

Carley, M. J., and E. O. Derow. 1983. <u>Social Impact Assessment: A Cross-Disciplinary Guide to the Literature</u>. Boulder, Colorado: Westview Press.

Carley, M. J., and E. S. Bustelo. 1984. <u>Social Impact Assessment: A Guide to the Literature</u>. Boulder, Colorado: Westview Press.

Carpenter, R. A., and J. E. Maragos. 1989. <u>How to Assess Environmental Impacts on Tropical Islands and Coastal Areas: A Training Manual</u>. Honolulu, Hawaii: East-West Center, Environment and Policy Institute.

Center for Environmental Management and Planning. 1986. <u>The EEC Environmental Assessment Directive: Towards Implementation</u>. Scotland: Aberdeen University and the United Kingdom Department of Environment.

Child, R. D. and others. 1987. <u>Arid and Semi-Arid Rangelands: Guides for Development</u>. Natural Resources Expanded Information Base Project. Morrelton, Arkansas: Winrock International.

Chironis, N. P., ed. 1980. <u>Training Manual for Miners</u>. New York: McGraw-Hill.

Clark, B. D. and others. 1981. <u>A Manual for the Assessment of Major Development Proposals</u>. London, United Kingdom: HMSO.

Clark, B. D., R. Bisset, and P. Wathern. 1980. <u>Environmental Impact Assessment: A Bibliography with Abstracts</u>. London, United Kingdom: Mansell Publishers.

Clark, B. D. and others, eds. 1984. <u>Perspectives on Environmental Impact Assessment</u>. Dordrecht, The Netherlands: Riedel.

Clark, M., and J. Herrington, eds. 1988. <u>The Role of Environmental Assessment in the Planning Process</u>. London, United Kingdom: Mansell Publishers.

Cohrrsen, J. J., and V. T. Covello. 1989. <u>Risk Analysis: A Guide to Principles and Methods for Analyzing Health and Environmental Risks</u>. Washington D.C.: Council on Environmental Quality.

Conway, G. 1986. <u>Agroecosystem Analysis for Research and Development</u>. Bangkok, Thailand: Winrock International Institute for Agricultural Development.

Conway, G. R., ed. 1986. The Assessment of Environmental Problems. London, United Kingdom: Imperial College, Centre for Environmental Technology.

Cook, P. L. 1983. A Review of the Recent Research on the Utility of Environmental Impact Assessment. Chania, Crete: Environmental Impact Assessment Symposium.

___. 1979. Costs of Environmental Impact Statements and the Benefits they Yield to Improvements in Projects and Opportunities for Public Involvement. Villach, Austria: Economic Commission for Europe.

Covello, V. T. and others, eds. 1985. Environmental Impact Assessment and Risk Analysis: Contributions from the Psychological and Decision Sciences. New York: Springer Publishing Company.

Daly, H. E., and J. B. Cobb. 1989. For the Common Good: Redirecting the Economy toward Community, the Environment, and a Sustainable Future. Boston, Massachusetts: Beacon Press.

Davies, G. S., and F. G. Muller. 1983. A Handbook on Environmental Assessment for Use in Developing Countries. Nairobi, Kenya: UNEP.

DeJongh, P. E. 1985. Environmental Impact Assessment: Methodologies, Prediction and Uncertainty. Utrecht, The Netherlands: IAIA Congress.

___. 1985. Technical Aspects of Training in Environmental Impact Assessment, with Emphasis on Ecological Impacts. Maastricht, The Netherlands: European Institute for Public Administration.

Derman, W., and S. Whiteford. 1985. Social Impact Analysis and Development Planning in the Third World. Boulder, Colorado: Westview Press.

Dixon, J. D. and others, eds. 1988. Economic Analysis of the Environmental Impacts of Development Projects. London/Manila: Earthscan Publications Ltd. and Asian Development Bank.

Draggan, S., J. J. Cohrsson, and R. E. Morrison, eds. 1987. Environmental Monitoring, Assessment and Management. New York: Praeger.

Duinker, P. N. 1989. "Ecological Efforts Monitoring in Environmental Impact Assessment: What Can It Accomplish?" Environmental Management 13:797-805.

Eberhardt, L. L. 1976. "Quantitative Ecology and Impact Assessment." Journal of Environmental Management 4:27-70.

Economic Commission for Europe. 1990. Post-Project Analysis for Environmental Impact Analysis. New York: United Nations.

Elkin, T. J., and P. G. R. Smith. 1988. "What is a Good Environmental Impact Statement? Reviewing Screening Reports from Canada's National Parks." Journal of Environmental Management 26(1): 71-89.

Elking-Savatsky, P. D. 1986. Differential Social Impacts of Rural Resource Development. Boulder, Colorado: Westview Press.

England, R. W., and E. P. Mitchell. 1990. "Federal Regulation and Environmental Impact of the U.S. Nuclear Power Industry, 1973-1984." Natural Resources Journal 30:537-539.

Environmental Resources Limited. 1990. Environmental Assessment Procedures in the U.N. System. A Study prepared at the request of the United Nations System. London, United Kingdom.

Erikson, P. A. 1979. Environmental Impact Assessment: Principles and Applications. New York: Academic Press.

Evans, J. 1982. Plantation Forestry in the Tropics. Oxford, United Kingdom: Clarendon Press.

Evers, F. W. R. 1986. "Environmental Assistance and Development Assistance: The Work of the OECD." In Methods and Experiences in Impact Assessment, edited by H. A. Becker and A. L. Porter. Atlanta, Georgia: International Association for Impact Assessment.

Finnish Department of International Development Cooperation. 1989. "Guidelines for Environmental Assessment in Development Assistance." (Draft). Finland: FINNIDA.

Finsterbusch, K., J. Ingersoll, and L. G. Llewellyn. 1990. Methods for Social Analysis in Developing Countries. Boulder, Colorado: Westview Press.

Finsterbusch, K., and C. P. Wolfe, eds. 1977. Methodology of Social Impact Assessment. Stroudsberg Pennsylvania: Dowden, Hutchinson and Ross.

Flavin, C. 1988. "The Case Against Reviving Nuclear Power." Worldwatch 1:27-35.

Food and Agricultural Organization. 1982. Environmental Impact Analysis and Agricultural Development. FAO Environment Paper 2. Rome, Italy.

___. 1982. "Environmental Impact of Forestry." Conservation Guide 7:1-85.

___. 1977c. Planning Forest Roads and Harvesting Systems. Forestry Paper 2. Rome, Italy.

Fortlage, C.A., 1990. Environmental Assessment: A Practical Guide. Aldershot, United Kingdom: Gower.

Frideres, J. S., and J. E. DiSanto, eds. 1986. Issues of Impact Assessment: Development of Natural Resources. Atlanta, Georgia: International Association for Impact Assessment.

Gamman, J. K., and S. T. McCreary. 1988. "Suggestions for Integrating Environmental Impact Assessment and Economic Development in the Caribbean." Environmental Impact Assessment Review 8(1):43-62.

Gas Research Institute. 1988. Environmental Aspects of Rights-of-Way for Natural Gas Transmission Pipelines: An Updated Bibliography. Prepared by the National Laboratory, Energy and Environmental Systems Division. Argonne, Illinois.

Gehrisch, W. and others. 1989. "The Potential Longer Term Contribution of Nuclear Energy in Reducing CO_2 Emissions in OECD Countries." Paris OECD/IEA Symposium: Energy Technologies for Reducing Emissions of Greenhouse Gases 1:619-634.

Glenn, J. C. 1988. Livestock Production in North Africa and the Middle East. Problems and Perspectives. World Bank Discussion Paper 38. Washington, D.C.: World Bank.

Go, F.C. 1987. Environmental Impact Assessment: An Analysis of the Methodological and Substantive Issues Affecting Human Health Considerations. London, United Kingdom: Monitoring and Assessment Research Centre/WHO/UNEP.

___. 1988. Environmental Impact Assessment: Operational Cost Benefit Analysis. London, United Kingdom: King's College, Monitoring and Assessment Research Center.

Goldberg, E. D., ed. 1982. Atmospheric Chemistry. Berlin: Springer Verlag.

Golden, J. and others. 1979. Environmental Impact Data Book. Ann Arbor, Michigan: Ann Arbor Science.

Gooden, P. M., and A. I. Johnstone. 1988. "Environmental Impact Assessment: Its Potential Application to Appropriate Technology in Developing Countries." The Environmentalist 8(1): 57-66.

Goodland, R. 1989. "The Environmental Implications of Major Projects in Third World Development. In Major Projects and the Environment, edited by P. Morris. Oxford, United Kingdom: Major Projects Association.

___. ed. 1990. The Race to Save the Tropics. Washington, D.C.: Island Press.

Goodland, R., C. Watson, and G. Ledec. 1985. Environmental Management in Tropical Agriculture. Boulder, Colorado: Westview Press.

Gorse, J. E., and D. R. Steeds. 1988. Desertification in the Sahelian and Sudanian Zones of West Africa. World Bank Technical Paper 61. Washington, D.C.: World Bank.

Gough, J. D. 1989. <u>Strategic Approach to the Use of Environmental Impact Assessment and Risk Assessment Within the Decision-Making Process</u>. Center for Resource Management Paper 13. New Zealand: University of Canterbury and Lincoln College.

Gunnerson, C. G., and D. C. Stuckey. 1986. <u>Anaerobic Digestion: Principles and Practices for Biogas Systems</u>. World Bank Technical Paper 49. Washington, D.C.: World Bank.

Gunnerson, C. G. 1989. <u>Post-Audits of Environmental Programs and Projects</u>. New York: American Society of Civil Engineers.

Hall, A. L., and J. Midgley, eds. 1988. <u>Development Policies: Sociological Perspectives</u>. Manchester, United Kingdom: Manchester University Press.

Highton, N. H., and M. Y. Chadwick. 1982. "The Effects of Changing Patterns of Energy Use on Sulfur Emissions and Depositions in Europe." <u>Ambio</u> 11:324-329.

Hipel, K. W. 1988. "Nonsparametric Approaches to Environmental Impact Assessment." <u>Water Resource Bulletin</u> 24(3):487-492.

Holling, C. S., ed. 1978. <u>Adaptive Environmental Assessment and Management</u>. New York: John Wiley and Sons.

Horberry, J. A. J. 1984. "Development Assistance and the Environment." Ph.D. Dissertation, Massachusetts Institute of Technology. Cambridge, Massachusetts.

___. 1987. "Environmental Impact Assessment for Development." <u>ATAS Bulletin</u> 4:59-60.

Horstmann, K., comp. 1985. <u>Environmental Impact Assessment for Development</u>, ed. K. Klennert. Feldafing, Federal Republic of Germany.

Howe, G. M., ed. 1977. <u>A World Geography of Human Diseases</u>. New York: Academic Press.

Hufschmidt, M. M., and R. A. Carpenter. 1980. <u>Natural Systems Assessment and Benefit-Cost Analysis for Economic Development</u>. Honolulu, Hawaii: East-West Center.

Hufschmidt, M. M. and others. 1983. <u>Environment, Natural Systems and Development: An Economic Development Guide</u>. Baltimore, Maryland: The John Hopkins University.

Hunsaker, C. T. and others. 1990. "Assessing Ecological Risk on a Regional Scale." <u>Environmental Management</u> 14:325-332.

Hyman, E., and B. Stiftel. 1988. <u>Combining Facts and Values in Environmental Impact Assessment: Theories and Techniques</u>. Boulder, Colorado: Westview Press.

Ingram, G. K. 1984. "Housing Demand in the Developing Metropolis: Estimates from Bogota and Cali, Colombia." World Bank Staff Working Paper 733. World Bank, Washington, D.C.

Institute for Solid Wastes of the American Public Works Association. 1975. Solid Waste Collection Practice. Chicago, Illinois.

Inter-American Development Bank. 1991. Application of Environmental Procedures in the Agricultural Sector. Guidelines. 2nd edition. Washington, D.C.

___. 1991. Application of Environmental Procedures in the Sanitation and Urban Development Sector. Guidelines. Washington, D.C.

___. 1990. Procedures for Classifying and Evaluating Environmental Impacts of Bank Operations. Washington, D.C.

International Commission for Environmental Assessment. 1990. Final Report of the Working Group on an International Commission for Environmental Assessment. Utrecht, The Netherlands.

Jaffe, L. S. 1973. "Carbon Monoxide in the Biosphere: Sources, Distributions and Concentrations." Journal of Geophysical Research 67(5):293-305.

Janicke, M. and others. 1989. "Economic Structure and Environmental Impacts: East-West Comparisons." Environmentalist 9:171-83.

Janikowski, R., and A. Starzewska. 1986. "Environmental Impact Assessment Project in Poland." Environmental Impact Worldletter (May-June):1-4.

Khosla, P. K., and D. K. Khurana, eds. 1987. Agroforestry for Rural Needs. New Delhi, India: Indian Society of Tree Scientists.

Khan, S. A. 1987. "Social Impact of Agricultural Development in Bangladesh: A critique of the Differentiation/Polarization Thesis." Journal of Social Studies 37:15-29.

Kneese, A. V., and J. L. Sweeney, eds. 1985. Handbook of Natural Resource and Energy Economics. 2 Volumes. New York: North-Holland.

Lal, R. and others. 1983. Land Clearing in the Tropics. Boston, Massachusetts: A. A. Balkema.

Lang, R., and A. Armour. 1980. Environmental Planning Resourcebook. Ottawa, Canada: Environment Canada, Lands Directorate.

Lavine, M. J. and others. 1978. "Bridging the Gap between Economic and Environmental Concerns in Environmental Impact Assessment." Environmental Impact Assessment Review 2. New York: Elsevier Science Publishing Company.

Lee, N. 1987. Environmental Impact Assessment: A Training Guide. Department of Town and Country Planning Paper 18. Manchester, United Kingdom: University of Manchester.

___. 1982. "The Future Development of Environmental Impact Assessment." Journal of Environmental Management 14:71-90.

___. 1984. Training for Environmental Impact Assessment. Brussels, Belgium: Economic Commission for Europe.

Lee, N., and C. M. Wood. 1985. "Training for Environmental Impact Assessment within the Economic Commission for Europe." Journal of Environmental Management 21:271-286.

Lee, N., C. M. Wood, and V. Gazidellis. 1985. Arrangements for Environmental Impact Assessment and Their Training Implications in the European Communities and North America. Department Department of Town and Country Planning Paper 13. Manchester, United Kingdom: University of Manchester.

Leistritz, F. L., and B. L. Ekstrom. 1986. Social Impact Assessment and Management: An Annotated Bibliography. New York: Garland Publishers.

Leistritz, F. L., and S. H. Murdock. 1981. The Socioeconomic Impact of Resource Development. Boulder, Colorado: Westview Press.

Leopold, L. B. and others. 1971. A Procedure for Evaluating Environmental Impact. U.S. Geological Survey Circular 645. Washington, D.C.

Levy, J. P., 1976. "Aplicacion Preliminar del Metodo de Evaluacion de Impacto Ambiental Ocasionado por la Planta Nucleo Electrica de Laguna Verde, Veracruz." Tesis Profesional, Universidad Nacional Autonoma de Mexico. Mexico: Facultad de Ciencias.

Lichfield, N. 1989. "Environmental Impact Assessment in Project Appraisal in Britain." Project Appraisal 3:133-141.

Logan, J. A. and others. 1981. "Tropospheric Chemistry: A Global Perspective." Journal of Geophysical Research 86(7):210-254.

Lohani, B. N. 1984. Environmental Quality Management. New Delhi, India: South Asian Publishers.

Macrory, R., and M. Lafontaine. 1982. Public Enquiry and Enquete Publique. London, United Kingdom: Institute for European Environmental Policy.

Metcalf & Eddy, Inc. 1972. Wastewater Engineering: Collection, Treatment, Disposal. New York: McGraw-Hill.

McCormick, J. 1985. Acid Earth: A Global Threat of Acid Pollution. Washington, D.C.: International Institute for Environment and Development.

Ofori-Cudjoe, S. 1990. "Environmental Impact Assessment in Ghana. An Ex-Post Evaluation of the Volta Resettlement Scheme: The Case of the Kpong Hydro-Electric Project." Environmentalist 10:115-126.

___. 1991. "Environmental Impact Assessment in Ghana: Current Administration and Procedures Towards an Appropriate Methodology." Environmentalist 11:45-54.

Opschoor, H. and D. Pearce, eds. 1991. Persistent Pollutants: Economics and Policy. Boston, Massachusetts: Kluwer Academic Press.

O'Riordan, T., and W. R. D. Sewell, eds. 1981. Project Appraisal and Policy Review. Chichester, United Kingdom: John Wiley and Sons.

O'Riordan, T. 1989. "The Impact of Environmental Impact Assessment on Decision-Making." In Environmental Impact Assessment, edited by V. T. Covello. Heidelburg, Federal Republic of Germany: Springer.

Pearce, D. W., E. Barbier, and A. Markandya. 1990. Sustainable Development: Economics and Environment in the Third World. Aldershot, United Kingdom: Elgar.

Pendse, Y. D., R. V. Rao, and P. K. Sharma. 1989. "Environmental Impact Assessment Methodologies: Shortcomings and Appropriateness for Water Resources Projects in Developing Countries." International Journal of Water Resources Development 5(4):252-259.

Pethig, R. and K. Fiedler. 1989. "Effluent Charges on Municipal Waste Water Treatment Facilities: In Search of Their Theoretical Rationale." Economics 49(1):71-74.

Peto, J. 1979. "Dose-Response Relationships for Asbestos-Related Disease: Implications for Hygiene Standards. Part II. Mortality." Annals of the New York Academy of Science 330:195-203.

Pimentel, D. 1989. "Agriculture and Ecotechnology." In Ecological Engineering: An Introduction to Ecotechnology, edited by W. J. Mitsch and S. E. Jorgensen. New York: John Wiley and Sons.

Porter, A. L. and others. 1980. A Guidebook for Technology Assessment and Impact Analysis. New York: North Holland.

Prieur, M. 1984. "Les Etudes d'Impact en Droit Francais." Zeitschrift fur Umweltpolitik 4:367-388.

Pryde, P. R. 1987. "The Soviet Approach to Environmental Impact Analysis." In Environmental Problems in the Soviet Union and Eastern Europe, edited by F. B. Singleton. Boulder, Colorado: Lynne Rienner Publishers.

Rau, J. G., and D. C. Wooten, eds. 1980. Environmental Impact Analysis Handbook. New York: McGraw-Hill.

Reynolds, L. 1981. "Foundations of an Institutional Theory of Regulation." Journal of Economic Issues 4:641-656.

Repetto, R., ed. 1985. The Global Possible Resources, Development and the New Century. New Haven, Connecticut: Yale University Press.

Roberts, R. D., and T. M. Roberts, eds. 1984. Planning and Ecology. London, United Kingdom: Chapman and Hall.

Ross, W. A. 1987. "Evaluating Environmental Impact Statements." Journal of Environmental Management 25(2):137-148.

Rossini, F. A., and A. L. Porter, eds. 1983. Integrated Impact Assessment. Boulder, Colorado: Westview Press.

Sadler, B., ed. 1987. Audit and Evaluation in Environmental Assessment and Management: Canadian and International Experience. 2 Volumes. Hull, Quebec: Environment Canada.

___. 1980. Public Participation in Environmental Decision Making: Strategies for Change. Edmonton, Alberta: Environment Council of Canada.

Sammy, G. K. 1982. "Environmental Impact in Developing Countries." Ph.D. Dissertation, University of Oklahoma. Norman, Oklahoma.

Schibuola, S., and P. H. Byer. 1991. "Use of Knowledge-Based Systems for the Review of Environmental Impact Assessments." Environmental Impact Assessment Review 11:11-27.

Seattle, D. M., and C. C. Patterson. 1980. "Lead in Albacore: Guide to Lead Pollution in Americans." Science 207(1):167-176.

Seidman, H., and I. J. Selikoff. 1990. "Decline in Death rates among Asbestos Insulation Workers, 1967-1986, Associated with Diminution of Work Exposure to Asbestos." Annals of the New York Academy of Science 609:300-318.

Shrader-Frechette, K. S. 1985. Science Policy, Ethics, and Economic Methodology: Some Problems of Technology Assessment and Environmental-Impact Analysis. Boston, Massachusetts: Riedel.

Sigal, L. L., and J. W. Webb. 1989. "The Programmatic Environmental Impact Statement: Its Purpose and Use." The Environmental Professional 11(1):14-17.

Sonntag, N. C. and others. 1987. Cumulative Effects Assessment: A Context for Further Research. Ottawa, Canada: Ministry of Services and Supply.

Stout, K. S. 1980. Mining Methods and Equipment. New York: McGraw-Hill.

Strickland, G. T. 1984. Hunter's Tropical Medicine. 6th edition. Philadelphia, Pennsylvania: W. B. Sanders.

Suter, G. W. and others. 1987. "Treatment of Risk in Environmental Impact Assessment." Environmental Management 11:295-303.

Tchobanoglous, G., H. Theisen, and R. Eliassen. 1977. Solid Wastes: Engineering Principles and Management Issues. New York: McGraw-Hill.

Tharun, G., N. C. Thanh, and R. Bidwell, eds. 1983. Environmental Management for Developing Countries. 3 Volumes. Bangkok, Thailand: Asian Institute of Technology.

Tidsell, C. 1986. "Cost-Benefit Analysis, the Environment and Informational Constraints in LDCs." Journal of Development 11:63-81.

Tomlinson, P., ed. 1987. "Environmental Audits: Special Edition." Environmental Monitoring and Assessment 8(3):183-261.

United Nations. 1990. Environmental Assessment Procedures in the UN System. London, United Kingdom: Environmental Resources Limited.

United Nations Environment Programme. 1982. The Use of Environmental Impact Assessment for Development Project Planning in ASEAN Countries. Bangkok, Thailand: Regional Office for Asia and the Pacific.

United Nations Environment Programme and the World Health Organization. 1989. Assessment of Urban Air Quality. London, United Kingdom: Global Environment Monitoring System.

United Nations and the United Nations Asian and Pacific Development Institute. 1980. Environmental Impact Statements: A Test Model Presentation, comp. C. Suriyakumaran. Bangkok, Thailand.

United States Council on Environmental Quality [and] Fish and Wildlife Service. 1980. Biological Evaluation of Environmental Impacts. Report FWS/OBS-80/26. Washington, D.C.

United States Department of Energy. 1986. Digest of Environmental and Water Statistics No. 9. Washington, D.C.: General Printing Office.

Vahter, V., ed. 1982. UNEP/WHO Assessment of Human Exposure to Lead and Cadmium through Biological Monitoring. Stockholm, Sweden: National Swedish Institute of Environmental Medicine and Karolinska Institute.

Vighi, M. and D. Calamari. 1990. "Evaluative Models and Field Work in Estimating Pesticide Exposure," in L. Sommerville and C. Walker, eds., Pesticides and Wildlife: Field Testing. London, United Kingdom: Taylor and Francis.

Vizayakumar, K. and Mohapatra, P. K. J. 1991. "Framework for Environmental Impact Analysis with Special Reference to India." Environmental Management 15:357-68.

Vlachos, E. 1990. "Assessing Long Range Cumulative Impacts." In Environmental Impact Assessment, edited by V. T. Covello. Heidelburg, Federal Republic of Germany: Springer.

Walters, C. 1986. Adaptive Management of Renewable Resources. New York: Macmillan.

Ware, G. W. 1980. "Effects of Pesticides on Nontarget Organisms." Residue Reviews 76:173-201.

Wandesforde-Smith, G., and I. Moreira. 1985. "Subnational Government and Environmental Impact Assessment in the Developing World: Bureaucratic Strategy and Political Change in Rio de Janeiro." Brazilian Environmental Impact Assessment Review 5:223-238.

Ward, D. V. 1978. Biological Environmental Impact Studies: Their Theory and Methods. New York: Academic Press.

Warner, M. L., and E. H. Preston. 1974. Review of Environmental Impact Assessment Methodologies. Washington D.C.: United States Environmental Protection Agency.

Wathern, P. 1984. "Methods for Assessing Indirect Impacts." In Perspectives on Environmental Impact Assessment, edited by B. D. Clark and others. Dordrecht, The Netherlands: Riedel.

Wathern P. and others. 1987. "Assessing the Environmental Impacts of Policy: A Generalized Framework for Appraisal." Landscape and Urban Planning 14:321-330.

Wathern, P., ed. 1988. Environmental Impact Assessment: Theory and Practice. London, United Kingdom: Unwin.

Wenger, R. B., W. Huadong, and Ma Xiaoying. 1990. "Environmental Impact Assessment in the People's Republic of China." Environmental Management 14:429-439.

Westman, W. E. 1985. Ecology, Impact Assessment and Environmental Planning. New York: John Wiley and Sons.

Wetsone, G. S., and A. Rosencranz. 1984. Acid Rain in Europe and North America. Washington, D.C.: Environmental Law Institute.

Williams, H. J. 1987. "Issues in the Control and Disposal of Hazardous Materials," in M. Chatterji, ed., Hazardous Materials Disposal. Avebury-Gower, Aldershot, pp. 59-70.

Wilson, D. G., ed. 1977. Handbook of Solid Waste Management. New York: Van Nostrand Reinhold Company.

Wood, C. M., and V. Cazidellis. 1985. <u>A Guide to Training Materials for Environmental Impact Assessment</u>. Department of Town and Country Planning Paper 14. Manchester, United Kingdom: University of Manchester.

World Bank. 1991. <u>Country Capacity to Conduct Environmental Assessments in Sub-Saharan Africa</u>. Technical Department, Africa Region. Environmental Division Working Paper 1. World Bank, Washington, D.C.

___. 1990. <u>Environmental Health Components for Water Supply, Sanitation and Urban Projects</u>. World Bank Technical Paper 121. Washington, D.C.: World Bank.

___. 1985. <u>Environment, Health and Safety Guidelines for Use of Hazardous Materials in Small and Medium Scale Industries</u>. Environment Department. Washington, D.C.: World Bank.

___. 1988. <u>Environmental Industrial Waste Control Guidelines</u>. Environment Department. Washington, D.C.: World Bank.

World Health Organization. 1974. <u>Health Project Management: A Manual of Procedures for Formulating and Implementing Health Projects</u>. Geneva, Switzerland.

Yates, E. D. 1989. <u>Environmental Impact Assessment: What it is and Why International Development Organizations Need it</u>. Washington, D.C.: Council on Environmental Quality.

Young, K., ed. 1988. <u>Women and Economic Development: Local, Regional and National Planning Strategies</u>. New York: Berg (for Oxford and United Nations Educational, Scientific and Cultural Organization).

Ziyun, F. 1989. "Environmental Impact Assessment of Yangtze Valley Projects." <u>International Water Power and Dam Construction</u> 41:36-39.

Distributors of World Bank Publications

ARGENTINA
Carlos Hirsch, SRL
Galería Guemes
Florida 165, 4th Floor-Ofc. 453/465
1333 Buenos Aires

AUSTRALIA, PAPUA NEW GUINEA, FIJI, SOLOMON ISLANDS, VANUATU, AND WESTERN SAMOA
D.A. Books & Journals
648 Whitehorse Road
Mitcham 3132
Victoria

AUSTRIA
Gerold and Co.
Graben 31
A-1011 Wien

BAHRAIN
Bahrain Research and Consultancy
 Associates Ltd.
P.O. Box 22103
Manama Town 317

BANGLADESH
Micro Industries Development
 Assistance Society (MIDAS)
House 5, Road 16
Dhanmondi R/Area
Dhaka 1209

 Branch offices:
 Main Road
 Maijdee Court
 Noakhali - 3800

 76, K.D.A. Avenue
 Kulna

BELGIUM
Jean De Lannoy
Av. du Roi 202
1060 Brussels

CANADA
Le Diffuseur
C.P. 85, 1501B rue Ampère
Boucherville, Québec
J4B 5E6

CHINA
China Financial & Economic
 Publishing House
8, Da Fo Si Dong Jie
Beijing

COLOMBIA
Infoenlace Ltda.
Apartado Aereo 34270
Bogota D.E.

COTE D'IVOIRE
Centre d'Edition et de Diffusion
 Africaines (CEDA)
04 B.P. 541
Abidjan 04 Plateau

CYPRUS
MEMRB Information Services
P.O. Box 2098
Nicosia

DENMARK
SamfundsLitteratur
Rosenoerns Allé 11
DK-1970 Frederiksberg C

DOMINICAN REPUBLIC
Editora Taller, C. por A.
Restauración e Isabel la Católica 309
Apartado Postal 2190
Santo Domingo

EL SALVADOR
Fusades
Avenida Manuel Enrique Araujo #3530
Edificio SISA, ler. Piso
San Salvador

EGYPT, ARAB REPUBLIC OF
Al Ahram
Al Galaa Street
Cairo

The Middle East Observer
8 Chawarbi Street
Cairo

FINLAND
Akateeminen Kirjakauppa
P.O. Box 128
SF-00101
Helsinki 10

FRANCE
World Bank Publications
66, avenue d'Iéna
75116 Paris

GERMANY
UNO-Verlag
Poppelsdorfer Allee 55
D-5300 Bonn 1

GREECE
KEME
24, Ippodamou Street Platia Plastiras
Athens-11635

GUATEMALA
Librerias Piedra Santa
5a. Calle 7-55
Zona 1
Guatemala City

HONG KONG, MACAO
Asia 2000 Ltd.
6 Fl., 146 Prince Edward
 Road, W.
Kowloon
Hong Kong

INDIA
Allied Publishers Private Ltd.
751 Mount Road
Madras - 600 002

 Branch offices:
 15 J.N. Heredia Marg
 Ballard Estate
 Bombay - 400 038

 13/14 Asaf Ali Road
 New Delhi - 110 002

 17 Chittaranjan Avenue
 Calcutta - 700 072

 Jayadeva Hostel Building
 5th Main Road Gandhinagar
 Bangalore - 560 009

 3-5-1129 Kachiguda Cross Road
 Hyderabad - 500 027

 Prarthana Flats, 2nd Floor
 Near Thakore Baug, Navrangpura
 Ahmedabad - 380 009

 Patiala House
 16-A Ashok Marg
 Lucknow - 226 001

INDONESIA
Pt. Indira Limited
Jl. Sam Ratulangi 37
P.O. Box 181
Jakarta Pusat

ITALY
Licosa Commissionaria Sansoni SPA
Via Benedetto Fortini, 120/10
Casella Postale 552
50125 Florence

JAPAN
Eastern Book Service
37-3, Hongo 3-Chome, Bunkyo-ku 113
Tokyo

KENYA
Africa Book Service (E.A.) Ltd.
P.O. Box 45245
Nairobi

KOREA, REPUBLIC OF
Pan Korea Book Corporation
P.O. Box 101, Kwangwhamun
Seoul

KUWAIT
MEMRB Information Services
P.O. Box 5465

MALAYSIA
University of Malaya Cooperative
 Bookshop, Limited
P.O. Box 1127, Jalan Pantai Baru
Kuala Lumpur

MEXICO
INFOTEC
Apartado Postal 22-860
14060 Tlalpan, Mexico D.F.

MOROCCO
Société d'Etudes Marketing Marocaine
12 rue Mozart, Bd. d'Anfa
Casablanca

NETHERLANDS
InOr-Publikaties b.v.
P.O. Box 14
7240 BA Lochem

NEW ZEALAND
Hills Library and Information Service
Private Bag
New Market
Auckland

NIGERIA
University Press Limited
Three Crowns Building Jericho
Private Mail Bag 5095
Ibadan

NORWAY
Narvesen Information Center
Book Department
P.O. Box 6125 Etterstad
N-0602 Oslo 6

OMAN
MEMRB Information Services
P.O. Box 1613, Seeb Airport
Muscat

PAKISTAN
Mirza Book Agency
65, Shahrah-e-Quaid-e-Azam
P.O. Box No. 729
Lahore 3

PERU
Editorial Desarrollo SA
Apartado 3824
Lima

PHILIPPINES
International Book Center
Fifth Floor, Filipinas Life Building
Ayala Avenue, Makati
Metro Manila

POLAND
ORPAN
Palac Kultury i Nauki
00-901 Warszawa

PORTUGAL
Livraria Portugal
Rua Do Carmo 70-74
1200 Lisbon

SAUDI ARABIA, QATAR
Jarir Book Store
P.O. Box 3196
Riyadh 11471

MEMRB Information Services
 Branch offices:
 Al Alsa Street
 Al Dahna Center
 First Floor
 P.O. Box 7188
 Riyadh

 Haji Abdullah Alireza Building
 King Khaled Street
 P.O. Box 3969
 Dammam

 33, Mohammed Hassan Awad Street
 P.O. Box 5978
 Jeddah

SINGAPORE, TAIWAN, MYANMAR, BRUNEI
Information Publications
 Private, Ltd.
02-06 1st Fl., Pei-Fu Industrial
 Bldg.
24 New Industrial Road
Singapore 1953

SOUTH AFRICA, BOTSWANA
For single titles:
Oxford University Press
 Southern Africa
P.O. Box 1141
Cape Town 8000

For subscription orders:
International Subscription Service
P.O. Box 41095
Craighall
Johannesburg 2024

SPAIN
Mundi-Prensa Libros, S.A.
Castello 37
28001 Madrid

Libreria Internacional AEDOS
Consell de Cent, 391
08009 Barcelona

SRI LANKA AND THE MALDIVES
Lake House Bookshop
P.O. Box 244
100, Sir Chittampalam A.
 Gardiner Mawatha
Colombo 2

SWEDEN
For single titles:
Fritzes Fackboksforetaget
Regeringsgatan 12, Box 16356
S-103 27 Stockholm

For subscription orders:
Wennergren-Williams AB
Box 30004
S-104 25 Stockholm

SWITZERLAND
For single titles:
Librairie Payot
6, rue Grenus
Case postale 381
CH 1211 Geneva 11

For subscription orders:
Librairie Payot
Service des Abonnements
Case postale 3312
CH 1002 Lausanne

TANZANIA
Oxford University Press
P.O. Box 5299
Dar es Salaam

THAILAND
Central Department Store
306 Silom Road
Bangkok

TRINIDAD & TOBAGO, ANTIGUA BARBUDA, BARBADOS, DOMINICA, GRENADA, GUYANA, JAMAICA, MONTSERRAT, ST. KITTS & NEVIS, ST. LUCIA, ST. VINCENT & GRENADINES
Systematics Studies Unit
#9 Watts Street
Curepe
Trinidad, West Indies

UNITED ARAB EMIRATES
MEMRB Gulf Co.
P.O. Box 6097
Sharjah

UNITED KINGDOM
Microinfo Ltd.
P.O. Box 3
Alton, Hampshire GU34 2PG
England

VENEZUELA
Libreria del Este
Aptdo. 60.337
Caracas 1060-A

YUGOSLAVIA
Jugoslovenska Knjiga
P.O. Box 36
Trg Republike
YU-11000 Belgrade